INTRODUCTION TO MONOPULSE

Introduction to monopulse

DONALD R. RHODES, Ph.D.

University Professor,
North Carolina State University

ARTECH HOUSE, INC. • DEDHAM, MA

Preface

IT HAS BEEN SAID that monopulse means different things to different people. That this is so is understandable in view of the bewildering number of techniques that can properly be called monopulse. To some who have become absorbed in the details of one particular manifestation of the concept, and to others who may have only a casual acquaintance with it, monopulse may well mean that particular system with which they happen to be familiar. Even to those who are familiar with more than one, the various forms may appear to be independent and unrelated. It has been felt that the obstacle to a complete understanding of the concept lies in a lack of definition. Once monopulse has been defined properly a consistent theory can be developed showing the underlying relationships among all of the various forms, and the apparent differences between them can then be resolved. Such a definition is proposed here for the first time in the form of three postulates. It evolved as a part of the unified theory of monopulse, developed during the winter of 1956–1957 and described in Chap. 2.

This monograph is the first general treatment of the concept of monopulse. It was written to acquaint the student and practicing engineer with the principles of monopulse, and hopefully to suggest new ideas for further development to those already actively engaged in the field. No specific design information is included, and relatively few experimental data. The emphasis is on concepts rather than specific systems. A few representative systems are included, however, to illustrate ways in which the basic concepts may be implemented. But they are by no means exhaustive, for the number of monopulse systems admitted by the unified theory is unlimited.

Some of the concepts presented here are new; many are not. Wherever possible a conscientious attempt has been made to credit the various inventions to their inventors, but, as in the case of many scientific areas developed by modern research team efforts, individual contributions frequently remain anonymous. With respect to the book itself I should like to express my sincere personal appreciation to Drs. J. Q. Brantley, Jr., and W. A. Flood for the many stimulating discussions that led to its being written; to G. M. Kirkpatrick, Dr. R. M. Page, and Sir Robert Watson-Watt for their encouragement and constructive criticism of the manuscript; and to the Cornell Aeronautical Laboratory, at which I was a member of the staff during preparation of most of the manuscript, for their sympathetic interest in and active sponsorship of its preparation. Finally, it is a particular pleasure to acknowledge the influence of Dr. Lillian R. Lieber in recognizing the importance of *attitude* in scientific thought.

The subject of monopulse as presented here is admittedly incomplete. The field itself is relatively new, having been developed primarily for military applications. Consequently much of it remains as classified security information that cannot as yet be published in the open literature. One particularly important area that has been almost completely neglected here is the effect of noise on monopulse angle accuracy. To wait until a complete treatment of the subject of monopulse could be presented would result in an indefinite delay in publication; hence it was decided to adopt the philosophy of Sir Robert Watson-Watt's famous "cult of the third best: The best never comes and the second best comes too late." It is hoped that this "third best" has come soon enough to help chart the future development of monopulse.

Donald R. Rhodes

Orlando, Fla.
January, 1958

Contents

1. Introduction to the monopulse concept

MONOPULSE is a concept of precision direction finding of a pulsed source of radiation. The direction of the pulsed source, such as a scattering target or radar beacon, is determined by comparing the signals received on two or more antenna patterns simultaneously. Although strictly a receiving concept, monopulse has been used primarily in the field of radar, where the pulse feature is utilized to determine range as well as angle of arrival. In the years since World War II it has become a highly useful and versatile addition to the art of radar, serving to augment conventional radar capabilities in some cases by increasing the precision with which angle measurements can be made, and in others by adding a third dimension to its data-gathering capabilities. The potential applications of monopulse are just beginning to be explored and should provide electronic engineers with new and fertile fields of activity for years to come.

The initial stimulus to develop monopulse was in the field of precision target tracking. Sequential-lobing techniques such as beam switching and conical scanning, used earlier for target tracking, were found to be degraded in accuracy by the effects of target scintillation. It was felt that a technique for determining direction by comparing the return on two or more antenna lobes *simultaneously* could be designed to eliminate this source of error. Furthermore, simultaneous lobing would have the advantage of a higher data rate since complete three-dimensional information on the target location would be available from every pulse

1

received, in principle at least.   It was this possibility of locating a target completely from the return of a single pulse that led to the term "monopulse," suggested originally by H. T. Budenbom[2,*] at the Bell Telephone Laboratories in 1946.

Simultaneous-lobing techniques were used to measure the angle of arrival in atmospheric propagation tests as early as 1928.[†] Within four years after the first operational radar tests in 1936 the British, under the direction of Sir Robert Watson-Watt, had an operating prototype of a search radar using a simultaneous-lobing technique for beam sharpening (Sec. 3.7).   In the same year, 1940, the first simultaneous-lobing experiments at the Naval Research Laboratory were performed.[‡]   The British and American work in this, as in so many other areas, were each unknown to the other, illustrating once again the amazing extent to which the two countries paralleled each other in the development of radar in all its forms.

These early experiments formed the background for the first work on monopulse in America.   During World War II the first organized efforts were generated within the laboratories of the General Electric Company,[1] the Naval Research Laboratory,[3] and the Bell Telephone Laboratories;[2] these efforts were directed primarily toward the improvement of target tracking.   Subsequently a number of other potential applications of the monopulse concept have appeared, branching out into many fields where precision angle-of-arrival measurements are required.

1.1   Sequential vs. simultaneous lobing

Various forms of automatic precision direction finding have been developed that are based either on sequential or on simultaneous lobing.   In either case the angle of arrival of the incoming wave is determined by comparing the signal received on two or more noncoincident antenna patterns.   The patterns are usually mirror images about an axis, called the *boresight* axis, of the antenna.   When the emitting source, which may be either a radio transmitter or a scattering target, is on the boresight axis, the signals received are equal.   Therefore the direction of a

* All numbered references are listed in the Bibliography.
† H. T. Friis, Oscillographic Observations on the Direction of Propagation and Fading of Short Waves, *Proc. IRE*, vol. 16, pp. 658–665, May, 1928.
‡ Private communication from Dr. R. M. Page.

given source may be determined by bringing the boresight axis into coincidence with the source direction as indicated by equal received signals. This is the physical basis of most target-tracking techniques, whether the comparison is made sequentially or simultaneously.

Direction finding by sequential lobing is performed by comparing the signals received by the various antenna patterns in a sequence. The most common techniques for sequential lobing are beam switching and conical scanning, on both of which a tremendous amount of work has been done. In the case of beam switching* the antenna patterns are produced in pairs, as indicated in Fig. 1.1. When a source appears within the two

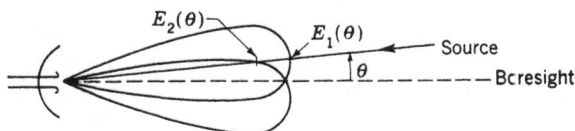

Fig. 1.1 Amplitude comparison of squinted beams.

beams at an angle θ from the boresight, a voltage of amplitude $E_1(\theta)$ is received by the upper beam and a voltage of amplitude $E_2(\theta)$ by the lower beam. By switching a receiver alternately between the two antenna feeds the two amplitudes of the received voltages may be compared. On the boresight axis they are equal. Above the boresight E_1 is greater than E_2, while below the boresight it is smaller. Thus the comparison provides an indication of the amount of angular displacement of the source from the boresight and, most importantly, the direction, or sense, of that displacement. Once the source has been acquired, it becomes possible to track it automatically, since any displacement of the source from the boresight results in an error signal that may be used to drive the antenna mechanically in a closed servo loop until that error is reduced to zero. By beam switching it is possible theoretically to obtain a new error signal on every pulse after the first. Comparing the return from each pulse with that received on the opposite beam from the preceding pulse indicates the average angular error of the source between the two pulses.

Rapid beam switching introduces a number of problems associated with the switching process itself. Furthermore, it is limited

* A. Maclese and J. Ashmead, The Rhumbatron Waveguide Switch, *J. IEE (London)*, vol. 93, pt. IIIA, no. 4, pp. 700–702, 1946.

in its simplest form to direction finding in a single plane; complete three-dimensional direction finding involves switching between two pairs of beams, the original pair plus another in a plane perpendicular to that of the first. Therefore three-dimensional direction finding by sequential lobing is usually performed by conical scanning.[4] When a single beam, such as one of those used for beam switching, is rotated about the boresight axis in a cone, the signal amplitude received from a source will be modulated at the conical scan frequency. The position of the modulation maxima corresponds to the direction of the source. As the boresight is brought onto the source, the percentage modulation will reduce to zero. The modulation amplitude indicates the amount of angular error, while the relative phase of the modulation envelope indicates its direction. As the source passes through the boresight axis, the modulation phase will reverse, indicating a reversal of sense. Therefore conical scanning provides all of the information needed for automatic tracking. It may be slower than beam switching, however, because the maximum rate at which error data can be collected is limited by the mechanical capability of the scanning element.

A severe limitation to the tracking accuracy of any sequential-lobing radar technique is the angular jitter caused by pulse-to-pulse fading of the return signal. One of the best records* published of radar fading is reproduced in Fig. 1.2. The measurements were made at a wavelength of 3 cm with a repetition rate of 1,500 pulses per second. When the radar beam was fixed in the direction of the target, the signal faded rapidly and at random from pulse to pulse (Fig. 1.2a). Fading of this nature is a result of target scintillation, a rapid change of radar cross section caused by a highly sensitive interference phenomenon among scattering elements of the target. When the same beam was misaligned from the scintillating target by scanning conically at a rate of 30 revolutions per second (rps), the signal was modulated as in Fig. 1.2b. The effect of rapid fading on conical scanning is to produce a serious degradation in tracking accuracy in the form of angular jitter of the misalignment signal. It can be smoothed by integrating over a sufficiently large number of scans, but this results in an undesirable lag in the tracking loop. If, in addition, the fading signal should happen to have a strong component near

* J. F. Coales, H. C. Calpine, and D. S. Watson, Naval Fire-control Radar, *J. IEE (London)*, vol. 93, pt. IIIA, no. 1, p. 324, 1946.

the scanning frequency, as in the case of propeller modulation in Fig. 1.2c, the misalignment signal contains components very near to the conical scan frequency which can be removed only by integrating over a period of several seconds. Thus the effects of fading on sequential lobing can be very serious indeed.

Simultaneous lobing, too, is performed by comparing antenna patterns, but it is distinguished from sequential lobing by the fact that the angle of arrival can be measured instantaneously and continuously by comparing the antenna patterns simultaneously. Consequently those errors in sequential-lobing measurements caused by fading of the received signals need not be present,

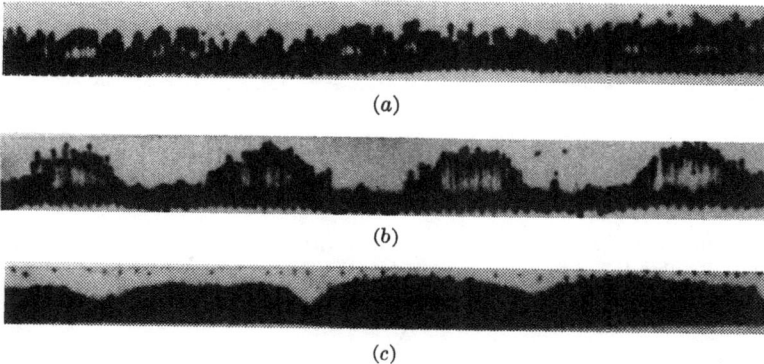

(a)

(b)

(c)

Fig. 1.2 Pulse-to-pulse fading of 3 cm radar signals: (a) target scintillation, (b) conical scanning at 30 cps, (c) fading with a strong 30-cps component caused by propeller modulation. (*From Coales, Calpine, and Watson.*)

since only the instantaneous *relative* amplitudes of the signals received from a given direction are measured, and these remain invariant. This is an important advantage. By pulsing the transmitted signal the instantaneous angle of arrival of radar return from such complicated scatterers as earth terrain can be measured in elevation (although not in azimuth), as discussed in Sec. 1.2. Such a measurement is based on the monopulse concept; it would be impossible using any other known radar technique.

In contrast to the various sequential-lobing techniques for obtaining angle of arrival, simultaneous lobing is not limited to a comparison of amplitudes alone. Phase patterns as well as amplitude patterns may be compared to obtain angle of arrival. Sequential lobing, on the other hand, is limited strictly to amplitude comparison unless special techniques are introduced for storing and comparing phase information. The angle of arrival, as obtained from the amplitude patterns in Fig. 1.1 by sequential comparison, may be obtained equally well by simultaneous comparison. But angle of arrival may be obtained instead from a phase-sensing antenna in the form of a radio interferometer whose phase patterns are compared simultaneously. Under certain conditions (Sec. 5.3) the phase patterns reduce to the patterns of a pair of displaced phase centers whose phase difference is a measure of angle of arrival. If the phase centers are spaced a

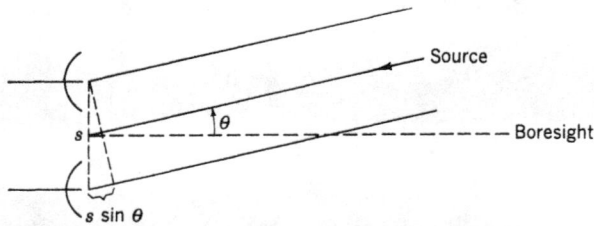

Fig. 1.3   Phase comparison by a radio interferometer.

distance s (Fig. 1.3), then the phase difference is related to angle of arrival θ by

$$\phi = \frac{2\pi}{\lambda} s \sin \theta. \tag{1.1}$$

It will be seen later in Chap. 5 that considerable flexibility of both amplitude- and phase-sensing characteristics is possible by controlling the aperture distribution of the antenna, although most modern simultaneous-lobing techniques have been limited either to the use of conventional radar beams squinted off of the boresight or to the use of displaced-phase-center interferometers.

The advantages of simultaneous lobing are obtained only at the expense of added complexity of the system. Instead of the usual single-channel receiver consisting of a mixer, intermediate-frequency (i-f) amplifier, and detector, it becomes necessary, for continuous angle measurement, to use at least two channels (for angle measurements of pulsed point sources a single channel may sometimes be shared[6]). The angle information is obtained by comparing the signals from each pair of channels. To ensure an accurate and stable angle indication these channels may have stringent dynamic requirements on amplitude or phase, or more generally on both amplitude and phase, since the comparison can, and always does in practice, involve the complex patterns of a simultaneous-lobing antenna.

1.2 The monopulse concept

The concept of monopulse emerged from the earlier concept of direction finding by simultaneous lobing. When an incoming wave from an isolated pulsed source at a sufficiently great range is received on two different antenna patterns simultaneously, the absolute amplitudes and absolute phases of the received signals may vary with the changing characteristics of the source or of the propagation medium, but their *relative* values are functions only of angle of arrival.[2] By focusing attention on the *ratio* of the pattern functions rather than on the pattern functions themselves, all of the parameters except angle of arrival are removed, in principle. The *instantaneous* formation of this ratio upon reception of each pulse to obtain the angle of arrival of radiation from all sources within the beam independently of their absolute amplitude levels is one of the two distinguishing characteristics of the monopulse concept. The other is symmetry of the angle output about the boresight axis.

In its simplest form the concept of monopulse involves a comparison of just a single pair of signals. This is sufficient to determine angle of arrival in a single plane, e.g., azimuth or elevation. Complete three-dimensional tracking, however, requires angle-of-arrival measurements in two orthogonal planes. Therefore nearly all tracking systems in use[1,3,4,7] involve comparisons of two pairs of signals, usually one pair in azimuth and the other in elevation.

In contrast to conical-scanning radar, in which the beam scans on both transmission and reception, the angle of arrival measured by a monopulse radar is determined solely by its receiving characteristics. Transmission remains the same as for any conventional pulse radar. This gives monopulse radar some other distinct advantages in addition to that of instantaneous precision direction finding, among them being the absence of any telltale scanning modulation and an increase in transmitted power in the boresight direction.[1]

In practice the angle of arrival determined by application of the concept of monopulse can never be measured instantaneously since there will always be some averaging present. In any

practical system the time constants of the system can never quite
be reduced to zero.   In some cases smoothing may be introduced deliberately in order to remove the random fluctuations in
angle of arrival caused by glint.   In an air-to-air radar application the Telecommunications Research Establishment found
that some smoothing had to be introduced into the system because
the point of apparent or real reflection from an aircraft appeared
to jump from wing to wing and to the tail.*

Boresighting vs. angle-scanning monopulse

The monopulse concept generally is applied in one of two ways,
regardless of whether it is used to sense the direction of an active
or of a passive radiating source.   In one way the output of the
system is taken directly as a measure of the angle of arrival of
the incoming wave relative to the boresight direction, while in
the other it is limited solely to an indication of the boresight
direction itself.   The latter application, boresighting, was the
first to be developed because it is the simplest, and since it is
also the most common it is sometimes mistaken to be the only
application.   It was developed originally for automatic precision
target tracking.   Once the target has been acquired, it can be
isolated from other targets and tracked in range by gating its
return, after which it can be tracked in angle by driving the
antenna from the angle output of the radar in a closed servo
loop.[1,6,7]   The other type of application that has been suggested,
and one that has some important implications, is instantaneous
angle scanning over the full width of the beam.   It has been
pointed out that "the entire sky could be scanned with one
revolution of the antenna.   Information on range, azimuth,
and elevation bearings, for all airplanes within the range of the
system, would be plotted on the oscilloscope each time the
antenna mount made one revolution."[1]

Angle of arrival

Existence of a well-defined angle of arrival is implied in the
concept of monopulse by the assumption that the pulsed source is
isolated in space at a sufficiently great range.   In the presence
of a number of isolated pulsed sources the angle of arrival from
each will still be well defined, provided no two are received
simultaneously.   Pulses received simultaneously from two simi-

* Private communication from Sir Robert Watson-Watt.

lar sources cannot be separated; hence the measured angle of arrival does not represent the direction of either source. When used to sense passive sources with a radar, where the pulsed sources are the scattering targets in the field of the radar transmitter, this means that the angle of arrival will be well defined providing no two radar targets at different angles within the beam lie within one pulse length radially of each other.

Even in the case of extended surface targets* a fairly well-defined angle of arrival may still be produced. The earth at near grazing incidence is an important example of such a target.

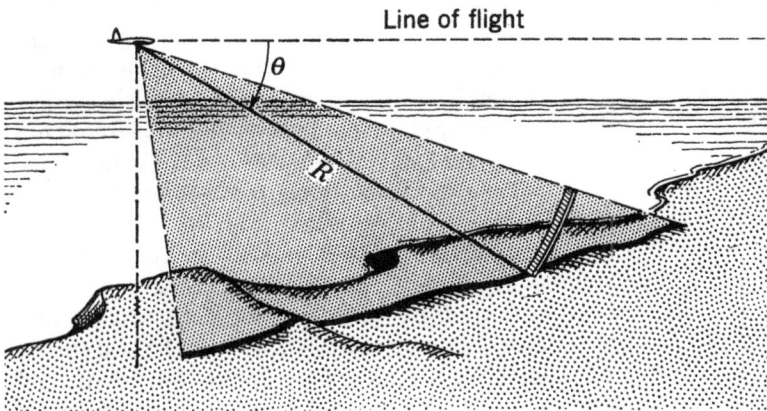

Fig. 1.4 Radar illumination of earth terrain by a narrow fan beam. The instantaneous angle of arrival θ is a continuous function of range R to the strip of terrain illuminated by the outward moving pulse.

If the radar beam is confined to a narrow fan shape in azimuth (Fig. 1.4), the radar return will originate from within the narrow strip of earth illuminated by the transmitted pulse. This element of the earth's surface will appear nearly as a point source since its dimensions are small relative to its range. As the pulse moves radially outward, the apparent point source will change in such a way as to move with it along the surface of the earth. The power scattered back at any time R/c after emission of the pulse will then arrive at a reasonably well-defined angle θ, an angle of arrival which will almost always be a single-valued function of time, and hence of range R, since the elevation profiles of terrain almost never double back on themselves. This is not true of

* L. N. Ridenour, "Radar System Engineering," Radiation Laboratory Series, vol. 1, McGraw-Hill Book Company, Inc., New York, 1947, p. 85.

most azimuth profiles of terrain at a given elevation, which is the reason why only the elevation profile of terrain may be described by the return from each pulse and hence why the beam must be narrow in azimuth. The angle of arrival will not be a set of discrete values, as in the case of truly isolated point targets, but instead it will vary continuously over the elevation profile as the pulse moves radially outward.

Angular resolution

Angular resolution, or the ability to distinguish the diffraction patterns of two point sources, is limited in conventional receiving systems to the half-power beamwidth of the diffraction pattern. This is the famous Rayleigh criterion of optics,* which applies equally well to conventional radio-frequency (r-f) systems. Monopulse, on the other hand, makes use of the range resolution inherent to the concept to relax this limitation. Any number of isolated pulsed sources, such as radar targets at different ranges, can be separated by gating (Fig. 1.5), after which their angular positions within the diffraction patterns of the antenna can be determined with considerably higher accuracy by simultaneous lobing. An increase in angular resolution by at least a factor of ten over the Rayleigh resolution is common. But without range resolution to isolate the individual pulsed sources within the beam, the angular resolution of monopulse reduces to exactly that of a conventional receiving system.

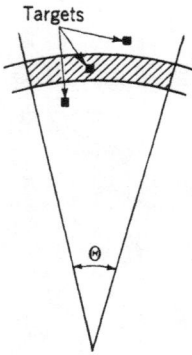

Fig. 1.5 Separation of radar targets by range gating (gated area of beam is shaded).

1.3   Angle noise

The effect of noise on any measurement of angle of arrival is to produce an uncertainty, or jitter, in the angular measurement. This was described for conical scanning in connection with the pulse-to-pulse fading records in Fig. 1.2. Monopulse, as well, is limited in angular accuracy by noise. It has an important advantage over any other form of pulsed direction finding, how-

* F. A. Jenkins and H. E. White, "Fundamentals of Physical Optics," McGraw-Hill Book Company, Inc., New York, 1937, p. 120.

ever, in that it reduces considerably the errors caused by fading. It was to achieve this advantage that monopulse was first developed.

Angle noise can be caused by any or all of three distinctly different phenomena:* (1) fading, (2) glint, and (3) thermal noise. Only the latter, thermal noise, is absolutely irreducible. Its irreducible part is the thermal noise exchanged between the antenna and its surroundings.† By using high-gain r-f amplifiers with low noise figures the internal noise generated within the receiver itself can be reduced to an arbitrarily low level, but the noise from the antenna remains undiminished. Of the other two types of noise, that associated with fading is removed, ideally, by the ratio action of monopulse. This is one of the primary features of monopulse, one considered to be of such importance that it has been elevated to the rank of a fundamental postulate of the unified theory of monopulse proposed in Chap. 2. In practice this type of noise cannot be removed completely, however, since an infinite bandwidth would be required to produce a truly instantaneous ratio, although it can be reduced below the level of the residual noise in the receiver. Its origin is principally scintillation of radar targets, i.e., those changes in radar cross section of complex multiple scatterers (such as aircraft) seen when the aspect angle changes. The third type of noise, glint, is caused by a random wandering of the center of radiation about complex multiple scatterers as the aspect angle changes. Both glint and scintillation result almost exclusively from the effects of scattering. When monopulse is used to sense an active source rather than a scattering target, in an environment free of reflections, neither of these sources of noise is present.

The two types of angle noise that cannot be removed by the ratio action of monopulse, namely glint and thermal noise, may be likened to the two types of amplitude noise present in the output of a conventional radar receiver, namely clutter and thermal noise, respectively. Glint in an angle measurement, just as

* C. E. Brockner, Angular Jitter in Conventional Conical-scanning, Automatic-tracking Radar Systems, *Proc. IRE*, vol. 39, pp. 51–55, January, 1951. Servo jitter is a fourth source of noise that affects the accuracy of a boresight-tracking type of system but is ignored here since it is not a property of monopulse itself.

† J. L. Lawson and G. E. Uhlenbeck, "Threshold Signals," Radiation Laboratory Series, vol. 24, McGraw-Hill Book Company, Inc., New York, 1950, p. 103.

clutter in an amplitude measurement, is produced solely by sources external to the receiver itself. Both glint and clutter occur only when the signal received is produced by scattering. Strictly speaking neither of these is truly noise since they both are believed to possess a certain degree of correlation over a number of pulses.* Thermal noise, on the other hand, is completely uncorrelated from pulse to pulse in either the case of a monopulse or a conventional radar receiver.

The root mean square (rms) angular error arising from glint is inversely proportional to range, since glint is a parameter associated with a linear dimension of the scattering target, whereas that arising from thermal noise is directly proportional to the square of the range, since it is proportional to the noise-to-signal voltage ratio.† Therefore, although glint may be the predominant source of noise at close range, thermal noise establishes the ultimate and final upper bound on angular accuracy (or lower bound on angular error). Budenbom has made a study[2] of the thermal bound and published some data obtained from monopulse angle-tracking experiments performed at Bell Telephone Laboratories in 1948. While the results are not conclusive, they do support the contention that the lower bound on angular error is proportional to the square of the range.

1.4 Early monopulse radar systems

It is appropriate to conclude this introduction with a review of some of the first attempts to achieve monopulse operation. Three basically different radar techniques tried during World War II, representing each of the three classes of monopulse (Sec. 2.4), will be described here.

Phase comparison

One of the first monopulse radar systems, and conceptually one of the simplest, was built and tested by Blewett, Hansen, Troell, and Kirkpatrick at the General Electric Company.[1] It was basically a dual-plane tracking interferometer radar with phase-sensing receiving antennas, linear i-f amplifiers, and a conventional phase comparator for detecting the two signals at the output. The circuit is illustrated schematically in Fig. 1.6.

* *Ibid.*, p. 124.
† Brockner, *loc. cit.*

The antenna, shown mounted on its rotating pedestal in Fig. 1.7, consisted of four 16-in. parabolic reflectors cut down and welded together, each with a separate feed at its focal point. One reflector was used to transmit and the other three to receive, two in azimuth and two in elevation with one reflector common to both planes. In this way the duplexing problem was avoided. Each pair of receiving antennas constituted an interferometer with phase centers spaced about 15 in. apart, corresponding to an electrical spacing of about 12λ at the operating wavelength of 3.2 cm. From Eq. (1.1) the beamwidth between first nulls of the interference pattern was about $4.8°$, corresponding to a range of phase of -180 to $+180°$. The half-power beamwidth of the

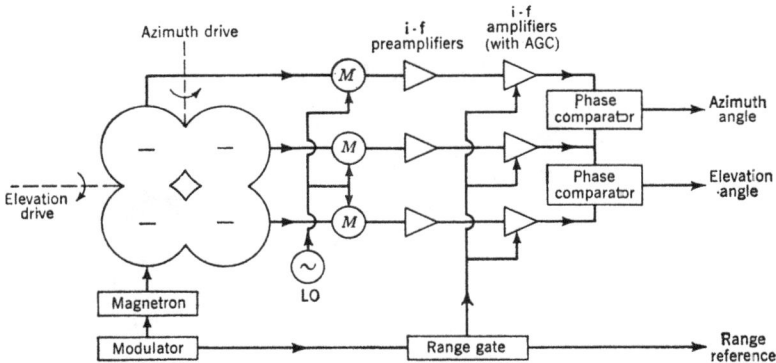

Fig. 1.6 An early dual-plane phase-comparison monopulse radar tracking system. (*After Blewett, Hansen, Troell, and Kirkpatrick.*)

paraboloids was somewhat greater, about $6°$, resulting in about $0.6°$ of reverse polarity signal from the adjacent interference lobes at the edges of the beam. Reverse polarity signals would be expected to cause a tracking radar to veer away from the target, and although no trouble was experienced it was recognized that an optimum interferometer design should have individual pattern beamwidths less than the composite interference beamwidth. If this system were to be used for angle scanning within the beam instead of boresight tracking, the pattern beamwidths should be limited to half of the interference beamwidth, because the output of the phase comparator used for angle detection is a multiple-valued function of angle of arrival for any range of phase exceeding -90 to $+90°$.

By heterodyning the r-f signals against a common local

oscillator they were converted to an intermediate frequency with
their relative phases as well as amplitudes preserved.   This is
essential to the operation of any phase-comparison system.   The
mixers, local oscillator, and i-f preamplifiers were mounted

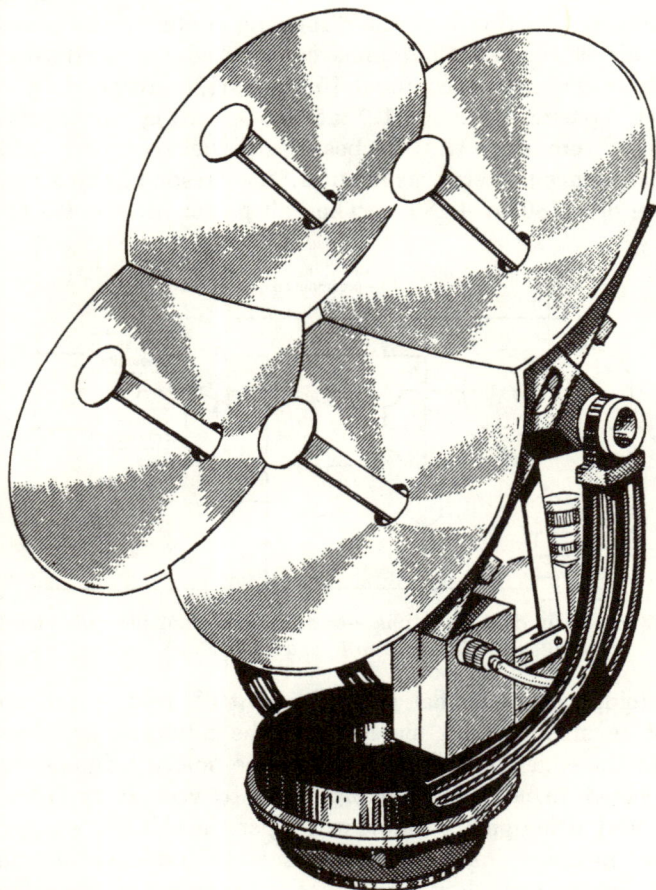

Fig. 1.7 The transmitting and receiving antennas of the General Electric
phase-comparison monopulse radar system, shown mounted on their
rotating pedestal.   (*General Electric Company.*)

directly on the back of the servo-driven clover-leaf antenna (Fig.
1.7) with all other components mounted on a stationary plat-
form.   Flexible cables were used to carry the i-f signals to the
i-f amplifiers in order to avoid the problems associated with

multiple-channel rotary joints. This limited the antenna rotation to 180°. An ordinary single-channel rotary joint was used to carry the transmitted signal from the magnetron to the transmitting paraboloid. Standard 30-Mc radar i-f amplifiers with 2-Mc bandwidth were used. Automatic gain control (AGC) signals were fed back to the grids of the second and fourth i-f amplifier stages to eliminate amplitude variations in the i-f output. An automatic range gate, synchronized by the modulator, was applied to the screens of the first and third stages to track the target automatically in range. The two pairs of constant-amplitude range-gated i-f signals were then detected in phase comparators with square-law rectifiers to produce an azimuth and an elevation angular error signal. When the target was on the boresight axis, both error signals were zero. Above or to the right of the boresight the error signals were positive, while below or to the left they were negative. The indicated boresight direction was found to be very stable over long periods of time, varying less than 0.025° in continuous operation over a period of a week. After video amplification of the pulsed outputs from the phase comparators, the video pulses were smoothed and used to drive a motor to position the antenna array. The monopulse receiver and positioning drive constituted a closed-loop servomechanism. The smoothed output from the receiver went through a null and reversed polarity when a target passed from one side of the boresight to the other, thereby reversing the driving motor rotation. The net result was automatic three-dimensional tracking in range, azimuth, and elevation. Tests in tracking a 100-mph airplane at ranges of about 1,000 yd showed a probable error of about 0.07° in azimuth and 0.05° in elevation, corresponding roughly to an electrical angular error of 4 to 5° in the measurement of relative phase of the signals received by the interferometer antennas.

Amplitude comparison

Perhaps the first amplitude-comparison monopulse system to be developed was a single-plane tracking radar invented by Sommers.[6] The angle of arrival was sensed by a pair of amplitude patterns whose main beams were squinted off of the boresight by lateral displacement of dipole feeds in a parabolic reflector (Fig. 1.8). Separate transmitting and receiving antennas were used as in the previous system to avoid the duplexing problem, with the two reflectors linked together mechanically for angle tracking.

A particularly original feature of the Sommers system is its use of a single channel, made possible by utilizing a characteristic unique to amplitude comparison of signals received from isolated pulsed sources. The angle information returned from an isolated target is contained solely in the relative amplitudes of the signals received on the return pulse by the two feeds; hence the pulses received may both be passed through a common receiving channel (mixer, i-f amplifier, and rectifier), if one of them is delayed in time by at least its own length. The delay network in the test model consisted of 100 ft of coiled wave guide to give a

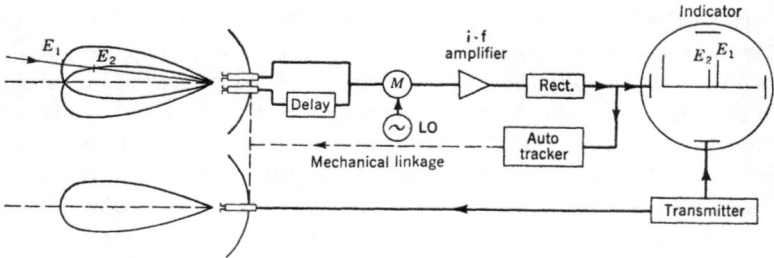

Fig. 1.8 An amplitude-comparison monopulse radar tracking system. The two receiving channels are combined by time sharing. (*After Sommers.*)

delay of one pulse length. With any other type of monopulse it would have been necessary to preserve phase as well as amplitude of the delayed signal. The use of a single channel is a tremendous simplification, eliminating much of the added complexity usually associated with monopulse. It is also an inherently balanced system since the channel appears nearly identical to the two pulses. After rectification the two video pulses may be separated by gating, and then they may be compared after delaying the first pulse by the same length of time that the second pulse had been delayed earlier. The two pulses may be compared independently of their absolute level simply by taking their difference, providing that a receiver is chosen whose transfer function has a logarithmic amplitude response (Sec. 3.4). Since the difference of two logarithms is the logarithm of their ratio, the two signals are normalized, in effect, one with respect to the other, just as they were in the previous system by the use of automatic gain control. On the boresight the two pulses are of equal height, hence the difference of their logarithms vanishes. About the boresight the difference of their logarithms exhibits odd symmetry. Although Sommers did not specify the characteristics of his receiver, this could have been the technique used to develop the error signal that drove the antenna assembly in a closed servo loop.

Two targets lying within a spherical shell two pulse lengths in thickness will have an overlapping interval in which the direct return from the far target and the delayed return from the near target will pass through the receiver simultaneously. Since the overlapping portions of the return pulses cannot be separated at the receiver output, it is evident that the Sommers system is limited to targets separated by a minimum radial distance of at least two pulse lengths. In most tracking applications this is not a serious limitation, however. Even with separate amplifying channels the minimum radial separation required would be about one pulse length. But in some other applications, particularly those in which continuous measurements of angle of arrival are necessary as in Fig. 1.4, the Sommers system of time-sharing a single channel is inapplicable.

Sum-and-difference comparison

A third form of monopulse, based on a comparison of the sum and the difference of the received signals, was developed by Page[3] at the Naval Research Laboratory. Whether the received signals are obtained from an amplitude- or a phase-sensing antenna, their difference is an odd function about the boresight axis and their sum an even function. The ratio of the difference to the sum is always independent of the absolute level of the received signals, in common with the other two systems described, and it bears a fixed relationship to the magnitude and sense of the angle of arrival relative to the boresight. As in the other two systems this ratio is formed, in effect, but in a rather unusual way, as will be seen.

The Page system is shown in Fig. 1.9 for sensing in a single plane. The angle information is obtained by amplitude sensing, although phase sensing could have been used just as well. An important difference between this and the other two systems will be observed in the use of a single antenna for both transmission and reception. Duplexing is relatively simple and straightforward whenever a sum signal is formed, for the sum pattern

bears a close resemblance to the type of pattern desired for transmission. The transmitter may then be connected to the sum
line during transmission by use of a transmit-receive (TR) switch
as was done here. The sum-and-difference signals were formed
by a hybrid ring, then amplified, and finally compared in a phase
comparator. The video output of the phase comparator was
proportional to the product of the sum-and-difference amplitudes. This product, $\Sigma \cdot \Delta$, was then added to an A-scope time-
base signal and compared on the A-scope with the square-law
rectified output Σ^2 of the sum channel. The result was a pip
that leaned to the right or to the left, depending on whether the
angle of arrival was on one side or the other of the boresight
direction. Precisely on the boresight the pip was vertical. The

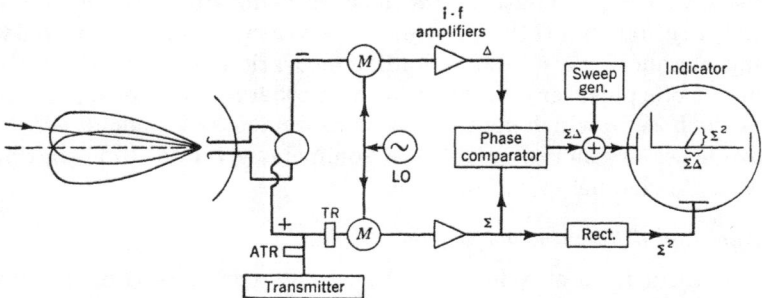

Fig. 1.9 An early sum-and-difference comparison monopulse radar with
duplexing. (*After Page.*)

magnitude of the pip varied with the strength of the received
signal, but its slope, $\tan^{-1} (\Sigma/\Delta)$, was independent of absolute
signal strength. Thus the slope of the pip indicated angle of
arrival independently of the absolute level through the ratio
Δ/Σ.

Bell Telephone Laboratories pioneered independently in the
use of the sum-and-difference method. Their development and
testing of sum-and-difference monopulse radar to improve tracking accuracy paralleled that of NRL's but was about a year later.
Neither laboratory was aware of the work being done at the other
until the earliest successful tests at NRL were disclosed by Dr.
Page to Mr. W. C. Tinus of BTL. At that time the Bell Telephone Laboratories were in the process of construction of their
first experimental model.* BTL conducted an analytical and

* Private communication from Dr. R. M. Page.

experimental study to compare a four-lens antenna with a four-horn cluster feeding a single-lens antenna.   This led finally to selection of the single lens for a tactical radar design based on the finding that its side-lobe level was lower.   Fractional-mil tracking results were obtained[2] with this system during the fall and winter of 1947–1948.*

* Private communication from Mr. R. A. Cushman.

2. A unified theory of monopulse

THE MONOPULSE concept has been described together with some of its properties in an effort to provide some insight into its significance. No precise definition of the concept has been given as yet, however, and consequently the various forms of monopulse considered so far may appear to the reader to be unrelated. It is proposed here that the concept be defined by three postulates. These three postulates lead to a completely general theory of monopulse. When restricted to the special cases of pure amplitude or pure phase sensing, the general theory reduces to a special theory that includes as particular cases not only the three described in Sec. 1.4 but *all* known physical manifestations of the concept. Thus the three postulates form the basis of a unified theory of monopulse. Development of the unified theory and exploration of its consequences are the principal thesis of this monograph.

In the development of the unified theory it should be borne in mind, and indeed cannot be emphasized strongly enough, that the concept of monopulse as defined by the three postulates is strictly an abstract idealization. No monopulse system has been or ever will be built that can satisfy this definition exactly. This should come as no surprise to the reader, however, since most physical theories represent just such abstractions. The importance of the theory developed here, as in many other physical theories, lies in the fact that it serves to tie together all the various forms of the concept, both those that have been known in the past and some that may not have been, into a simple and consistent framework. In so doing it serves to define the theoretical limits of practical monopulse systems.

A number of monopulse systems have been designed for applications where it was either unnecessary or disadvantageous to satisfy the three postulates below. According to this definition, then, such systems would not be considered true monopulse. Even in the case of such "pseudomonopulse" systems, however, many of the important characteristics derived from these postulates still apply, again emphasizing the importance of even an idealized theory.

2.1 Three postulates of monopulse

The source of monopulse angle information lies in the radiation patterns of the monopulse antenna. In general these may be complex. By comparing and processing the signals received, the angle information is extracted. The form in which it appears will depend on the way in which the signals are compared and processed, which in turn will determine the various types of monopulse that are possible. In order to establish a rigorous definition from which the essential characteristics of the various possible forms of monopulse may be deduced and developed into a unified theory, a set of three postulates will be hypothesized as the foundation of the theory. The postulates have not been chosen arbitrarily. They are explicit statements of the physical characteristics common to all monopulse systems, characteristics that are usually considered as being implicit to the concept of monopulse.

The three postulates are as follows:

Postulate 1: *Monopulse angle information appears in the form of a ratio.*

Angle information is sensed by comparing pairs of received signals. By requiring that the comparison be in the form of a ratio, the angle output of the monopulse system will be a function only of angle of arrival, independently of the absolute amplitude level of the signals received.

Postulate 2: *The sensing ratio for a positive angle of arrival is the inverse of the ratio for an equal negative angle.*

This postulate and the next impose conditions of symmetry about the boresight on the angle-output function. The term *inverse* is defined in the group sense,[*] i.e., any element times its inverse is the identity element.   When limited to the ordinary arithmetic group operations of multiplication and addition, only multiplicative and additive ratios are admissible as group elements.   If we denote the multiplicative and additive sensing ratios by $r_m(u)$ and $r_a(u)$, respectively, where u is itself a function of angle of arrival θ [Eq. (5.1)], then Postulate 2 states that

$$r_m(u) = \frac{1}{r_m(-u)}, \tag{2.1a}$$

and

$$r_a(u) = -r_a(-u). \tag{2.1b}$$

The multiplicative inverse of a group element is its reciprocal; its identity element is unity.   The additive inverse of a group element is its negative, its identity element being zero.   The physical interpretation of the identity element is that it represents the value of the sensing ratio in the boresight direction.

It is readily seen that the set of all sensing ratios, which may be the set of all complex numbers in general, with respect to either of the two arithmetic operations does indeed form a group; i.e., it satisfies the four postulates that define a group, namely, (1) the operation of any element in the set on any other element is also an element in the set, (2) successive operations are associative, (3) each element has an inverse, and (4) the set contains an identity element.   The group concept frequently penetrates to the basic structure of physical phenomena, for "wherever groups disclosed themselves, or could be introduced, simplicity crystallized out of comparative chaos."[†]

Postulate 3: *The angle-output function is an odd, real function of the angle of arrival.*

The angle output of a monopulse system must indicate magnitude and sense of the angle of arrival.   By this final postulate it is hypothesized that the angle output is a real function of angle of arrival and that it has odd symmetry about the boresight direction.

Angle output is a function of the sensing ratio $r(u)$.   With the exception of certain special but important cases $r(u)$ is complex,

* L. R. Lieber and H. G. Lieber, "Galois and the Theory of Groups," The Galois Institute of Mathematics and Art, Brooklyn, N. Y., 1956.

† E. T. Bell, "Mathematics, Queen and Servant of Science," McGraw-Hill Book Company, Inc., New York, 1951, p. 164.

being derived from antenna radiation patterns which are generally complex functions of angle. Therefore by this postulate the angle output will, in general, be expressed as a real function of the complex variable $r(u)$. If we define the angle output to be the real part of a complex angle-detection function $\mathfrak{F}(r)$, then Postulate 3 states that

$$\operatorname{Re} \mathfrak{F}[r(u)] = -\operatorname{Re} \mathfrak{F}[r(-u)]. \tag{2.2}$$

2.2 The general theory

It will be seen shortly that the operation of a monopulse system as defined by the three postulates may be described analytically by a sequence of mapping transformations of a function of a complex variable from one complex plane to another. The first of

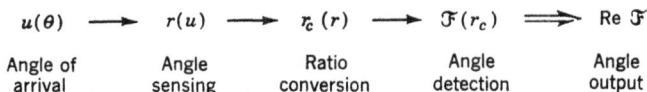

$$u(\theta) \longrightarrow r(u) \longrightarrow r_c(r) \longrightarrow \mathfrak{F}(r_c) \Longrightarrow \operatorname{Re} \mathfrak{F}$$

| Angle of arrival | Angle sensing | Ratio conversion | Angle detection | Angle output |

Fig. 2.1 Sequence of mapping transformations describing the operation of a monopulse system.

these transforms u, a fixed real function of angle of arrival θ, into the sensing ratio $r(u)$ of Postulate 2. This in turn may be converted by an appropriate transformation into any of the other forms of the sensing ratio admitted by the three postulates. Finally it is transformed into the angle detection function $\mathfrak{F}(r)$ of Postulate 3, the real part of which represents the angle output of the system. This sequence of transformations (Fig. 2.1) is the essence of the general theory of monopulse.

The sequence begins with the angle of arrival of the incoming wave. In any radiating system the full significance of the concept of pattern function can be seen only if the angle of arrival θ is interpreted as a complex variable (Sec. 5.1). The coordinate u then represents a map of the ϑ-plane contour onto the real axis of the complex $\pi(d/\lambda) \sin \vartheta$ plane. Its physical interpretation, shown by Woodward and Lawson in an analysis of diffraction through a hole in a conducting planar sheet, is that the usual visible range of real angles is associated with real power radiated into space, and that an additional range of complex angles must be included to account for reactive power appearing as energy stored about the aperture. The only ϑ-plane contour of visible and invisible angles permitted physically is the one that restricts $\pi(d/\lambda) \sin \vartheta$, whose real part is defined by*

$$u \triangleq \pi \frac{d}{\lambda} \sin \theta, \tag{2.3}$$

to the real axis.   Only that portion of the contour corresponding to the visible interval $-\pi d/\lambda \le u \le \pi d/\lambda$ may be associated physically with actual radiated power.   All the rest, corresponding to complex angles of arrival, must be associated with stored energy.

Angle sensing

The first transformation in the sequence characterizing the monopulse system is the formation of the complex angle-sensing ratio.   From the first and second postulates the sensing ratio will appear in one of the following forms:

$$r_m(u) = \frac{\mathcal{P}(u)}{\mathcal{P}(-u)} \tag{2.4a}$$

or
$$r_a(u) = \frac{\mathcal{P}_o(u)}{\mathcal{P}_e(u)}, \tag{2.4b}$$

where $\mathcal{P}(u)$ is an arbitrary function of u, and $\mathcal{P}_o(u)$ and $\mathcal{P}_e(u)$ are arbitrary odd and even functions of u, respectively.   For negative u the ratio $r_m(u)$ becomes its reciprocal and $r_a(u)$ becomes its negative, as required.   It remains to relate these functions to the antenna radiation patterns.   The simplest relationship possible for $r_m(u)$ is obtained if $\mathcal{P}(u)$ and $\mathcal{P}(-u)$ are taken to be the complex antenna patterns themselves.   The functions $\mathcal{P}_o(u)$ and $\mathcal{P}_e(u)$ can then be taken to be the odd and even components of $\mathcal{P}(u)$:

$$\begin{aligned} \mathcal{P}_o(u) &= \tfrac{1}{2}[\mathcal{P}(u) - \mathcal{P}(-u)] \\ \mathcal{P}_e(u) &= \tfrac{1}{2}[\mathcal{P}(u) + \mathcal{P}(-u)]. \end{aligned} \tag{2.5}$$

With this interpretation the ratios $r_m(u)$ and $r_a(u)$ are arbitrary but not independent.   Each can be converted into the other by the bilinear transformation

$$r_a(u) = \frac{r_m(u) - 1}{r_m(u) + 1}. \tag{2.6}$$

Another consequence of this interpretation is that the pattern

* The symbol $\triangleq$ means equal *by definition.*

functions must necessarily be mirror images about the boresight direction.

Although the functions $\mathcal{P}(u)$ and $\mathcal{P}(-u)$ have been interpreted as the antenna pattern functions with $\mathcal{P}_o(u)$ and $\mathcal{P}_e(u)$ as their odd and even parts, it should be pointed out that this was quite arbitrary. Instead, the functions $\mathcal{P}_o(u)$ and $\mathcal{P}_e(u)$ could have been interpreted just as logically as being the pattern functions, with $\mathcal{P}(u)$ and $\mathcal{P}(-u)$ as their mirror-image components. The final choice of interpretation was purely a matter of convenience; most monopulse antenna patterns in practice have mirror-image symmetry rather than odd and even symmetry, hence it is more convenient to express the characteristics of monopulse systems in terms of mirror-image patterns. In the event that odd and even patterns become more convenient, all of the ensuing relations may be interpreted in those terms by converting from r_m to r_a [Eq. (2.6)].

In general, and in all practical cases, the pattern of an antenna is a complex function of angle. It may be represented by

$$\mathcal{P}(u) \triangleq P(u)e^{j\varphi(u)}, \qquad (2.7)$$

where $P(u)$ and $\varphi(u)$ are the far-field amplitude and phase patterns, respectively. Since conventional communication systems are insensitive to the phase patterns of their antennas, the radiation pattern of an antenna is frequently considered as being synonymous with its amplitude pattern. Monopulse, on the contrary, may be vitally affected by the phase as well as the amplitude patterns. Hence the sensing ratio is, in general, a complex function of angle of arrival.

Actual pattern functions may always be considered as analytic functions of the complex variable $\pi(d/\lambda) \sin \vartheta$ since they are always single valued and differentiable.* Therefore the sensing ratio is also an analytic function. As a consequence, the entire theory of analytic functions may be applied to the general theory of monopulse. It is this powerful tool of analysis that may constitute the key to future development of the general theory.

Ratio conversion

Once the angle-sensing ratio has been formed, it can be converted into any other ratio satisfying the monopulse postulates

* E. T. Copson, "Theory of Functions of a Complex Variable," Oxford University Press, London, 1935, p. 136.

by another transformation, the second in the sequence describing the operation of a monopulse system. The most general analytic transformation with a one-to-one correspondence between all points in two simple complex planes has been shown* to be the homographic, or bilinear, transformation. The term simple is used to exclude multileaved Riemann surfaces. The general bilinear transformation is of the form

$$w = \frac{az + b}{cz + d}. \tag{2.8}$$

It has the property of mapping circles into circles. The complex constants a, b, c, and d may be determined by mapping any three noncoincident points of one circle into any three noncoincident points of another. One particular example of ratio conversion has already been seen in Eq. (2.6) relating $r_m(u)$ to $r_a(u)$. When the sensing ratio is detected directly, the conversion reduces simply to the identity transformation. Ratio conversion is of considerable importance to the theory of monopulse, for it offers a means of controlling the form of angle information from the angle sensor to the angle detector.

Angle detection

The third and final transformation in the sequence is the formation of the angle-detection function $\mathfrak{F}(r)$. The angle-detection function may be any arbitrary function whose real part is odd and continuous with continuous derivatives. The continuity condition on its real part, the angle output of the system, is the usual requirement for physical realizability. Continuity is assured if $\mathfrak{F}(r)$ is an analytic function, as it will be in all practical cases. Therefore $\mathfrak{F}(r)$ may be any analytic function whose real part is arbitrary in the upper half (or in the lower half) of the r plane and whose imaginary part is arbitrary everywhere. This is evident by observing that the real part in the lower half plane is determined uniquely by the choice of function in the upper half plane, or vice versa (Fig. 2.2). For the multiplicative ratio the upper half plane will be defined to include the real axis from -1 to $+1$, the lower half plane including the rest. Then for the additive ratio, related to r_m by Eq. (2.6), the upper half plane will include the negative real axis and the lower half plane the

* E. A. Guillemin, "The Mathematics of Circuit Analysis," John Wiley & Sons, New York, 1949, p. 365. Copson, *op. cit.*, p. 187.

positive real axis.   The general form of the angle-detection function is derived in the Appendix, where it is shown that it may be any odd analytic function of r_a, and that as a function of r_m it may be constructed of any arbitrary set of values of Re $\mathfrak{F}^{(n)}(1)$ for odd n and of Im $\mathfrak{F}^{(n)}(1)$ for all n.

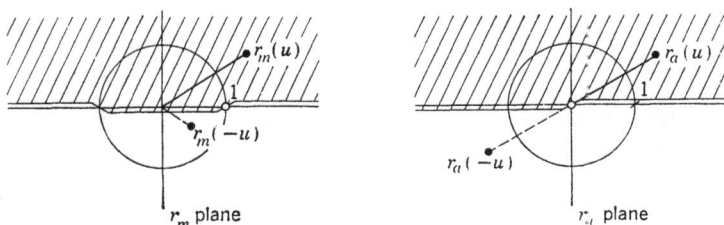

Fig. 2.2 Multiplicative and additive angle-sensing planes.   The boresight direction is indicated by the small circle o.

2.3   The special theory

Current monopulse practice is limited entirely to simple sensing of the angle of arrival, i.e., to either pure amplitude or pure phase sensing (in a single principal plane of the antenna).   But since all possible forms of simple sensing may be considered as special cases of the general theory, the special theory associated with them may be deduced from the general theory.   Therefore, the remainder of the monograph will be devoted to development of the special theory and to some of its consequences.

Consider first the case of pure amplitude sensing.   The angle information extracted by the monopulse antenna from the return wave is contained strictly in the amplitude patterns $P(u)$ and $P(-u)$ of the antenna.   The effective* phase patterns $\varphi(u)$ and $\varphi(-u)$ must be made identical with respect to a common origin.†   If the complex patterns have phase centers, as is fre-

---

* When the two received patterns are amplified directly (as described in Sec. 3.5), the nonsensing amplitudes or phases may be made equal *independently* of the antenna patterns; hence in this case the *effective* patterns of the nonsensing function are inherently equal.

† When taking the ratio of antenna patterns it is implied that the two patterns are referred to a common origin.   For amplitude patterns this is unimportant since far-field amplitude is insensitive to any finite displacement of the origin.   Phase patterns, on the contrary, are sensitive to displacements of even a small fraction of a wavelength.   Therefore it is essential that phase patterns always be referred to a common origin.

quently the case in practice (but not necessarily so as will be
seen in Sec. 5.3), any separation will introduce a corresponding
change in the angle measurement. With identical phase patterns
the two sensing ratios involve only pattern amplitudes:

$$r_m(u) = \frac{P(u)}{P(-u)},$$

and
$$r_a(u) = \frac{P(u) - P(-u)}{P(u) + P(-u)}.$$

Angle information is then contained solely in the amplitude ratio
$\rho(u)$, defined by

$$\rho(u) \triangleq \frac{P(u)}{P(-u)}; \tag{2.9}$$

hence
$$r_m(u) = \rho(u), \tag{2.10a}$$

and
$$r_a(u) = \frac{\rho(u) - 1}{\rho(u) + 1}. \tag{2.10b}$$

Considering next the case of pure phase sensing, the angle
information is contained strictly in the phase patterns $\varphi(u)$ and
$\varphi(-u)$. Here the effective amplitude patterns must be made
identical. With identical amplitude patterns the two sensing
ratios involve only pattern phases:

$$r_m(u) = \frac{e^{j\varphi(u)}}{e^{j\varphi(-u)}},$$

and
$$r_a(u) = \frac{e^{j\varphi(u)} - e^{j\varphi(-u)}}{e^{j\varphi(u)} + e^{j\varphi(-u)}}.$$

Angle information is then contained solely in the phase difference
$\phi(u)$, defined by

$$\phi(u) \triangleq \varphi(u) - \varphi(-u); \tag{2.11}$$

hence
$$r_m(u) = e^{j\phi(u)}, \tag{2.12a}$$

and
$$r_a(u) = j \tan \frac{\phi(u)}{2}. \tag{2.12b}$$

These four equations (2.10a,b and 2.12a,b) *constitute all the
various forms of pure amplitude or pure phase sensing permitted
by the monopulse postulates.* They describe a transformation of
the u axis into the four r-plane contours shown in Fig. 2.3. Pure
amplitude sensing is represented by a transformation to the real
axis of the r_m or r_a planes, pure phase sensing by a transformation to the unit circle in the r_m plane or to the imaginary axis in
the r_a plane.

The amplitude ratio $\rho(u)$ and phase difference $\phi(u)$ will be referred to as the *sensing functions*. More properly they are the multiplicative sensing functions. However, additive sensing can also be described in terms of these same functions by the transformations $[\rho(u) - 1]/[\rho(u) + 1]$ and $\tan [\phi(u)/2]$. Hence any type of sensor can be described by one or the other of the sensing functions.

The additive sensing functions, although not independent of the multiplicative sensing functions, nevertheless are of importance to the special theory in their own right. Their significance lies in the fact that they are both amplitude functions, one on the real axis and the other on the imaginary. Except for the

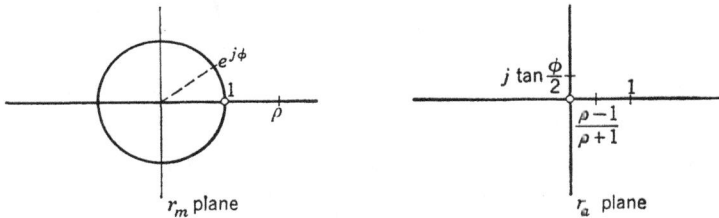

Fig. 2.3 The four r-plane contours admitted by the special theory.

90° phase difference between the two it would be impossible to tell whether amplitude or phase sensing had been used to obtain any given function. In this sense, therefore, the additive sensing functions are equivalent and hence *the sensing functions $\rho(u)$ and $\phi(u)$ are equivalent* themselves; i.e., the amplitude function $\rho(u)$ is equivalent to the phase function $\phi(u)$, and conversely, where the two are related by

$$\tan \frac{\phi(u)}{2} = \frac{\rho(u) - 1}{\rho(u) + 1}. \tag{2.13}$$

Equation (2.13) will be referred to as the *equivalence equation*. This equivalence between amplitude and phase sensing is of fundamental importance to the special theory, for it then becomes possible to divorce the concept of monopulse from any particular kind of sensor. Instead, any kind of sensor can be described in terms of either sensing function, or, more concisely if we wish, in terms of the sum-and-difference, or additive, sensing function $\Delta(u)/\Sigma(u)$:

$$\frac{\Delta(u)}{\Sigma(u)} \triangleq \begin{cases} \dfrac{\rho(u) - 1}{\rho(u) + 1} & \text{amplitude sensing,} \\[2ex] \tan \dfrac{\phi(u)}{2} & \text{phase sensing.} \end{cases} \tag{2.14}$$

For pure amplitude sensing $\Delta(u)/\Sigma(u)$ represents the first, and for pure phase sensing the second, since the two are equivalent. The equivalence between the amplitude- and phase-sensing functions is illustrated in Fig. 2.4, showing the correspondence between points in the complex plane. The right half of the unit circle corresponds to the positive real axis and the left half to the negative real axis. The symbols $\Delta(u)$ and $\Sigma(u)$ designate the magnitudes of the difference and of the sum, respectively, of

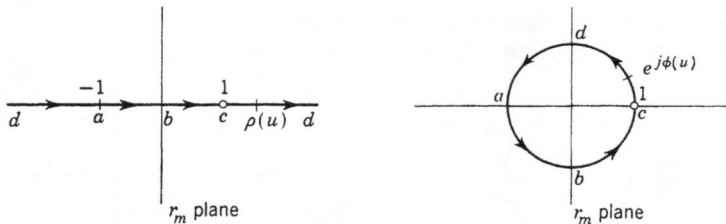

Fig. 2.4   Equivalence between the amplitude-sensing function $\rho(u)$ and the phase-sensing function $\phi(u)$.

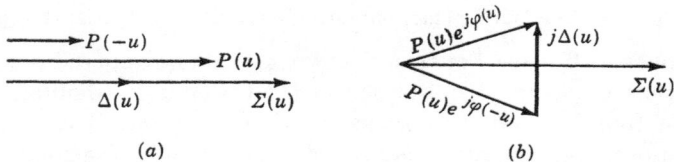

Fig. 2.5   Relationship between the complex-pattern functions and their sum and difference for (a) pure amplitude sensing, and (b) pure phase sensing.

the pattern functions. In the case of pure amplitude sensing the pattern functions are in-phase (Fig. 2.5a); hence their difference and sum are also in-phase:

$$\Delta(u) = P(u) - P(-u),$$
and
$$\Sigma(u) = P(u) + P(-u).$$

In the case of pure phase sensing the patterns have equal amplitudes (Fig. 2.5b); hence their difference and sum are in phase quadrature with magnitudes given by

$$\Delta(u) = 2P(u) \sin \frac{\phi(u)}{2},$$
and
$$\Sigma(u) = 2P(u) \cos \frac{\phi(u)}{2}.$$

The four forms of the sensing ratio admitted by the special theory are unique in that they are the only ones possible when limited solely to direct sensing. Twelve other forms are possible, however, by conversion of the original four into each of the other three. This would admit a total of sixteen possible forms. Eight of these are degenerate forms of the other eight, however, since the additive ratios have already been interpreted as conversions from the multiplicative ratios [Eq. (2.6)]. Furthermore, it will be seen that conversions between the two additive ratios are degenerate forms of each other, in view of the equivalence between sensing functions [Eq. (2.13)] and thence between the additive ratios themselves. This leaves a total of six nondegenerate forms of the sensing ratio possible, namely,

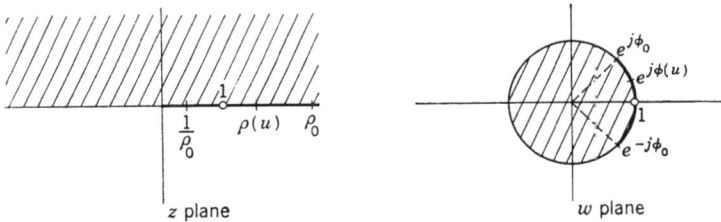

Fig. 2.6 Conversion between the amplitude- and phase-sensing functions.

the original four plus conversion of the two multiplicative forms into each other.

The two multiplicative conversions map the real axis onto the unit circle, and vice versa (Fig. 2.6). Circles are mapped into circles (the coordinate axes may be considered as circles of infinite radius), which can be expressed analytically by the general bilinear transformation [Eq. (2.8)]. In order that the boresight direction will remain unaltered by the conversion let the transformation map the point $r_m = 1$ into itself. The upper half plane may then be mapped into the interior of the unit circle in such a way that a portion of the real axis maps onto an arbitrary portion of the unit circle. In particular, to preserve symmetry the positive real axis must map into a symmetrical circular arc $-\phi_0 \leq \phi \leq \phi_0$, where ϕ_0 is arbitrary. The amplitude range equivalent to this range of phase is $\frac{1}{\rho_0} \leq \rho \leq \rho_0$, where ρ_0 is related to ϕ_0 by

$$\frac{\rho_0 - 1}{\rho_0 + 1} = \tan \frac{\phi_0}{2}. \tag{2.15}$$

The negative real axis then maps into the rest of the unit circle with the point -1 mapping into itself. The positive real axis in the r_m plane corresponds, on the u axis, to the interval about the boresight between first nulls of $P(u)$ and $P(-u)$. This corresponds, in turn, to an equivalent range of phase of $-90°$ $\leq \phi_{eq} \leq +90°$, by Eq. (2.13). Therefore by transforming the positive real axis into the unit circular arc $-\phi_0 \leq \phi \leq \phi_0$ we not only effect a conversion from amplitude to phase but also extend or compress the effective range of angle of arrival, depending on whether ϕ_0 is less than or greater than $90°$, respectively. The following three points determine the symmetrical mapping transformation uniquely:

$$z = 1 \quad \rightarrow \quad w = 1$$
$$z = 0 \quad \rightarrow \quad w = e^{-j\phi_0}$$
$$z = \infty \quad \rightarrow \quad w = e^{j\phi_0}.$$

Solving for the constants in Eq. (2.8), the transformation reduces to

$$w = \frac{e^{j\phi_0}z + 1}{z + e^{j\phi_0}}, \tag{2.16}$$

and since
$$z = \rho(u)$$
and
$$w = e^{j\phi(u)},$$

it becomes $\quad e^{j\phi(u)} = \exp j2 \tan^{-1}\left[\frac{\rho(u) - 1}{\rho(u) + 1} \tan \frac{\phi_0}{2}\right],$

or
$$\tan \frac{\phi(u)}{2} = \frac{\rho(u) - 1}{\rho(u) + 1} \tan \frac{\phi_0}{2}. \tag{2.17}$$

This is the *conversion equation* between the two sensing functions. It represents conversion either from amplitude to phase or from phase to amplitude, since the mapping transformation is bilinear. The factor $\tan (\phi_0/2)$ may be interpreted physically as an attenuation, or amplification, of the difference signal relative to the sum. Its effect is that of extending or compressing the realizable angular range. The realizable angular range is limited ultimately by the antenna patterns to something less than that corresponding to $-180° \leq \phi \leq 180°$, or its equivalent range of ρ, and may be limited further by the angle detector, usually to the range corresponding to $-90° \leq \phi \leq 90°$. But by proper choice of ϕ_0 it may be extended to the limit set by the antenna patterns, either by converting amplitude to phase with the difference voltage attenuated by a factor $\tan (\phi_0/2) < 1$, or by converting phase to amplitude with the difference voltage attenuated by a factor $\cot (\phi_0/2) < 1$. Circuits for realizing such conversions physically are illustrated in Sec. 3.3.

For the important special case where $\phi_0 = \pi/2$, the conversion equation reduces to the fundamental equivalence equation (2.13). It converts the amplitude-sensing function into the equivalent phase-sensing function, and vice versa. It is distinguished from the more general conversion equation by the fact that the angular range and boresight sensitivity associated with the antenna itself are preserved.

2.4 Classification of monopulse systems

The special theory limited the number of nondegenerate forms of the sensing ratio to a total of six. Three were associated with amplitude sensing and three with phase. It was concluded that there can be only three distinctly different forms of angle information to the angle detector in view of the fundamental equivalence between amplitude and phase sensing, because the angle detector cannot distinguish between actual and equivalent sensing functions. Thus there can be only three distinct kinds of angle detection, namely, amplitude, phase, and sum-and-difference. Either amplitude or phase sensing, which are equivalent to each other, may be used with each of the three. It is concluded further, then, that there can be only three distinct classes of special monopulse, the distinguishing characteristics of each being those of its angle detector.

The three classes of monopulse could be identified by the terms amplitude, phase, and sum-and-difference according to the type of angle detection. Each of these terms is also used to denote types of angle sensing, however, so it would be preferable to adopt a new notation. Therefore, they will be denoted simply as classes I, II, and III, where the numbers will be assigned in the order of the dynamic stability required of the system. In this way the nomenclature will serve to indicate one of the practical differences existing between the three classes. The principal sources of instability lie in the angle detector, for it is there that the active elements of the system usually are concentrated. Usually the angle sensor contains only passive elements; hence it is inherently stable and can be made free of any residual error by proper alignment. It is for this reason that the system classification can denote relative stability without reference to any particular angle sensor. The inherent stability of each of the three systems will now be examined.

Dynamic instability of the system transfer function can affect both its angle-scanning and its boresighting characteristics. In some applications the accuracy of the boresight alone is important, while in others both are important. Consider first the stability of the boresight. In the case of amplitude angle detection the boresight direction can be obtained by a comparison of amplitudes alone of the signals in the two channels. In the case of phase angle detection, however, the comparison can never be reduced to phases alone because normalization of amplitudes is never perfect, especially if the amplitudes fluctuate over a wide dynamic range. For sum-and-difference angle detection, on the other hand, the boresight is indicated not by equal signals in the two channels but rather by a null in the difference-channel amplitude. The location of a null is always totally unaffected by either amplitude or phase errors in the angle detector; hence we arrive at the important conclusion that its boresight indication will always be preserved exactly. Considering next the effect of dynamic instability on the shape of the angle-scanning angle-output function, it is again seen that with amplitude angle detection the angle-output function can be rendered insensitive to phase, whereas with phase angle detection in any practical system it will be affected more or less by both. In the case of sum-and-difference angle detection the angle information is contained solely in amplitude, but with sense. Phase per se is not involved. Thus the stability required of each of the three angle detectors may be summarized as follows:

Type of angle detection	Boresight stability depends on	Angle-output stability depends on
Amplitude	Amplitude	Amplitude
Phase	Amplitude and phase	Amplitude and phase
Sum-and-difference	Neither	Amplitude (with sense)

The order of classification is clear. Sum-and-difference angle detection is least sensitive to dynamic instability, in principle;

amplitude angle detection is next; phase angle detection is last. The three classes of monopulse are chosen, therefore, as follows: class I—sum-and-difference, class II—amplitude, and class III—phase.

Although the distinction between the two types of sensing was unnecessary in the classification of monopulse systems because of their fundamental equivalence, it will be a convenience to include such a distinction in the classification nomenclature. Therefore, it is suggested that the symbol A be added to denote amplitude and P to denote phase sensing. The six nondegenerate forms of monopulse can then be denoted as classes IA, IIA, IIIA, and IP, IIP, IIIP.

2.5 Illustration of the special theory

Use of the special theory to represent the operation of monopulse systems will be illustrated by treating a specific example. The actual systems represented will be described in Chap. 3, and will be used as illustrations throughout the remainder of this monograph. In this example the angle information will be obtained by amplitude sensing, although phase sensing could have been chosen just as well. The pattern functions chosen will be those generated by aperture distributions that are cosinusoidal in amplitude and linear in phase. The pattern functions will then be the same as they would be for an in-phase aperture distribution but shifted from the boresight axis $u = 0$ by the squint angle

$$u_s \triangleq \pi \frac{d}{\lambda} \sin \theta_s.$$

The pattern of an in-phase cosine distribution will be found to be proportional to the function

$$\frac{\cos u}{\frac{\pi^2}{4} - u^2}.$$

The optimum squint angle, at least for the case of class I (Sec. 6.3), is approximately half of the half-power beamwidth, or

$$u_s = 1.87.$$

If we choose this as the squint angle, the pattern functions are as

shown in Fig. 2.7a. The visible range of angle of arrival lies in
the interval $-\pi(d/\lambda) \leq u \leq \pi(d/\lambda)$. The invisible (dotted)
portion of the pattern lying outside of that interval is mani-
fested physically by reactive power stored about the aperture.
The sum and the difference of these pattern functions are shown
in Fig. 2.7b. The corresponding sensing function $\rho(u)$ representing the ratio of the pattern amplitudes, and the phase function
$\phi_{eq}(u)$ equivalent to it, are shown in Fig. 2.8a. An important
characteristic that may be observed in both the amplitude and
the equivalent phase functions is the usable range of angle scan,
seen here for this example to be about $-1.6\pi \leq u \leq 1.6\pi$.

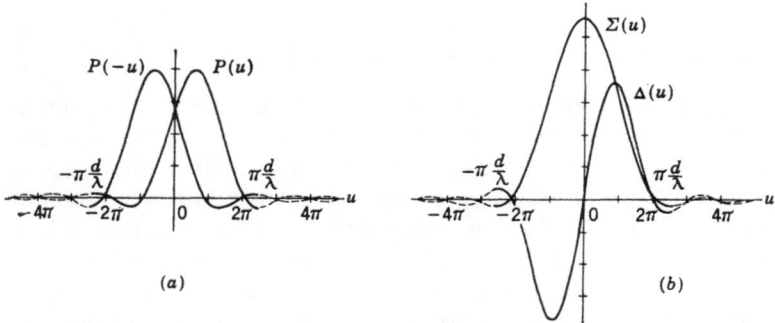

(a)   (b)

Fig. 2.7 Radiation characteristics of a monopulse amplitude-sensing
antenna with cosine aperture distributions and with beams squinted away
from the boresight by half of the beamwidth: (a) pattern functions, and (b)
the sum and the difference functions.

Beyond this range the angle output is ambiguous, no longer giving a single-valued indication of angle of arrival. The theoretical
upper bound on unambiguous angular range for any type of
monopulse system is limited inherently to $\pm 180°$, which is
seen to correspond to the beamwidth to first nulls of the sum
pattern. The multiplicative sensing ratios $\rho(u)$ and $e^{j\phi_{eq}(u)}$ over
the unambiguous angular range of this example are shown in
Fig. 2.8b. The positive real axis and the right half of the
unit circle correspond to the angular interval between first nulls
of the two pattern functions (Fig. 2.7a), which in turn corresponds roughly to the half-power beamwidth of the sum pattern
and to the region of positive slope of the difference pattern (Fig.
2.7b). Outside of this interval the sensing ratio involves side
lobes of one or both of the two pattern functions. If the side

lobes can be reduced below the minimum level of the fluctuating
signals received, a threshold level can be established to blank out
all indications of waves arriving at angles outside of the unambiguous angular range.   Only those arriving within the angular-
scanning range would then be admitted, and this without
ambiguity.

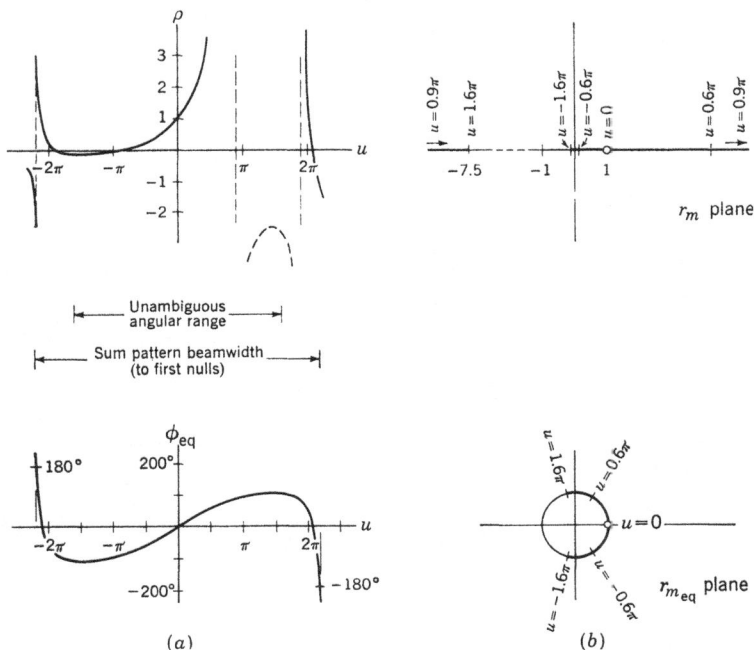

Fig. 2.8   Characteristics of the angle sensor: (a) the sensing function $\rho(u)$
and its equivalent $\phi_{eq}(u)$, and (b) the sensing ratio $r_m(u)$ and its equivalent
$r_{m_{eq}}(u)$.

So far only the angle sensor has been considered.   From that
alone some of the important characteristics of the monopulse
system have been observed independently of the choice of angle
detector.   Once the angle sensor has been chosen, it may be
used with any of the three classes of angle detection by making
an appropriate conversion of the sensing ratio.   After choosing
the particular form of angle detection for each class, a complete
analytical description can be expressed as a sequence of transformations between complex planes.   To complete this example
a typical angle-detection function for each of the three classes

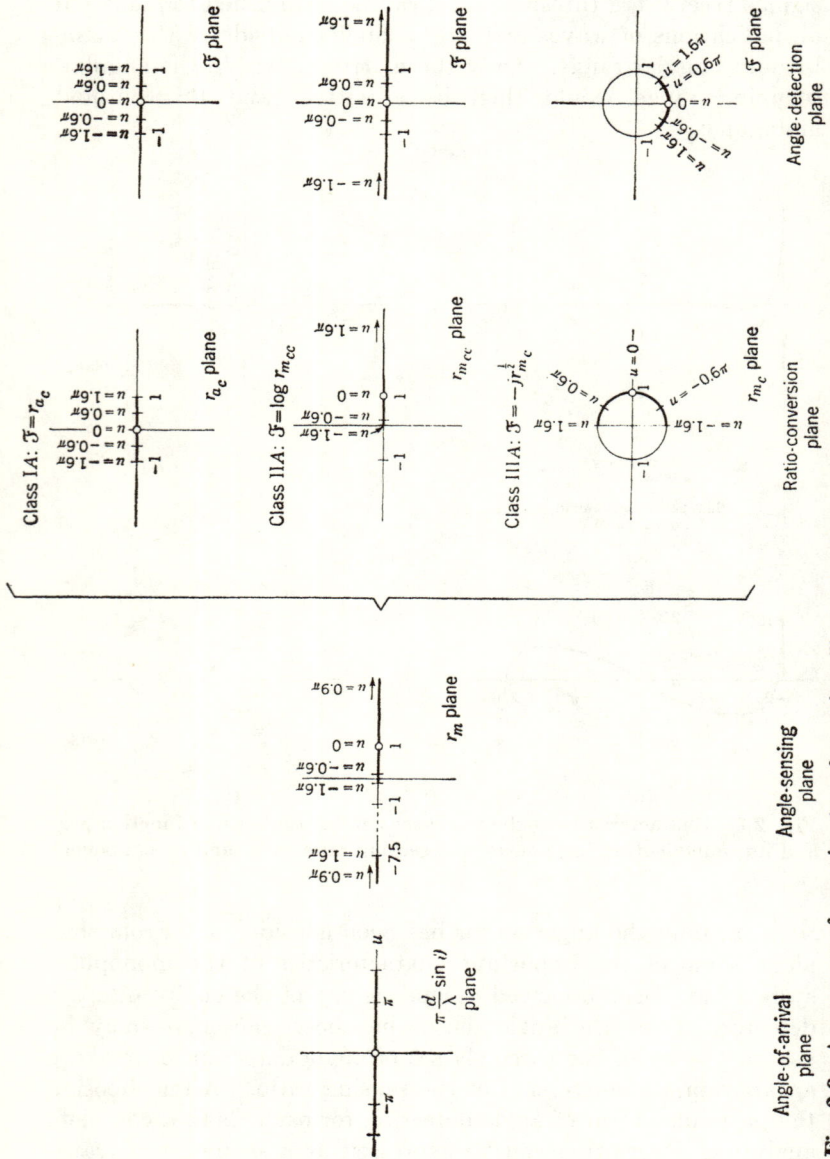

Fig. 2.9 A sequence of mapping transformations illustrating the operation of each of the three classes of monopulse.

will be chosen from among those of the angle detectors in common use today. Each is described briefly in Sec. 3.4. The functions chosen are as follows:

$$\mathfrak{F} = \begin{cases} r_{a_e} & \text{class I} \\ \log r_{m_{ee}} & \text{class II} \\ -jr_{m_e}^{1/2} & \text{class III.} \end{cases}$$

In each case the unambiguous range of angle detection is limited by the angle detector either to the positive real axis or to the right half of the unit circle, which are equivalent to each other by Eq. (2.13). But in each case the angular range may be extended to the maximum permitted by the angle sensor by performing the appropriate conversion transformation between the angle sensor and angle detector. The transformation required is that which maps the points $e^{\pm i\phi_0}$ into $e^{\pm i\pi/2}$, where ϕ_0 is the supplement of the maximum value of $\phi_{eq}(u)$. In this example the maximum value of $\phi_{eq}(u)$ is $105°$ (Fig. 2.8). The transformation is then effected by attenuating the difference voltage by a factor of $\tan \phi_0/2 = 0.77$.

The complete sequence of mapping transformations for all three classes is summarized in Fig. 2.9. The first transformation represents the formation of the angle-sensing ratio, a transformation involving only the angle sensor and hence one that is common to all three classes. The second is the ratio-conversion transformation, necessary in all three classes to extend the angular range but a different transformation for each of the three. The purpose of conversion in the case of class I is simply to extend its angular range; the transformation from amplitude to phase is immaterial since the amplitude and phase additive ratios have been found to be degenerate forms of each other. In the case of class II a double conversion is necessary (indicated by the double subscript), converting first from amplitude to phase in order to extend the range and then converting back again to amplitude. In the case of class III, conversion would have been necessary regardless of whether or not the range were to be extended since the amplitude-sensing ratio must be converted into an equivalent phase-sensing ratio. The converted sensing ratios for the three cases (doubly converted in the case of class II), when expressed in terms of the sum-and-difference sensing function $\Delta(u)/\Sigma(u)$, are as follows:

$$r_{a_c}(u) = \frac{\Delta(u)}{\Sigma(u)} \tan \frac{\phi_0}{2} \qquad\qquad \text{class I,}$$

$$r_{m_{cc}}(u) = \frac{1 + \dfrac{\Delta(u)}{\Sigma(u)} \tan \dfrac{\phi_0}{2}}{1 - \dfrac{\Delta(u)}{\Sigma(u)} \tan \dfrac{\phi_0}{2}} \qquad\qquad \text{class II,}$$

$$\text{and} \quad r_{m_c}(u) = \exp\left\{ j2 \tan^{-1}\left[\frac{\Delta(u)}{\Sigma(u)} \tan \frac{\phi_0}{2} \right] \right\} \qquad \text{class III.}$$

The third and final transformation in the sequence is the formation of the angle-detection function, whose real part is the angle output of the system.

For the three classes in this example the angle output obtained as a function of u is shown in Fig. 2.10. The effect of angular

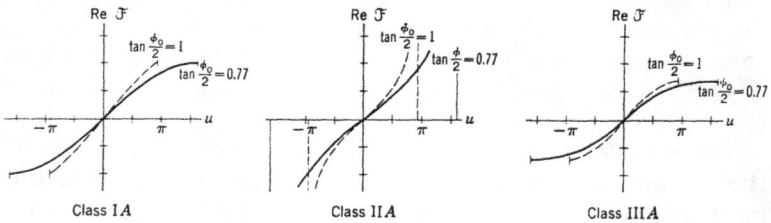

Fig. 2.10   Angle output of each of the three monopulse systems described in Fig. 2.9, with extension (solid) and without extension (dotted) of the angular range.

range extension by conversion of the sensing ratio may be seen directly. The solid lines show the angle output produced by attenuating the difference voltage for maximum unambiguous angular range, while the dotted lines show the angle output that would have resulted without angular-range extension. Thus the maximum angular range is obtained along with an increase in the angular range of approximate linearity, but at the expense of a decrease in boresight sensitivity. If sensitivity rather than angular range were the important parameter, as it might be in a boresighting application, it could be increased at the expense of reduced angular range by attenuating the sum signal to make $\tan (\phi_0/2)$ greater, rather than less, than unity. This exchange between angular range and sensitivity is a fundamental property of monopulse.

2.6 Summary

The concept of monopulse has been defined by a set of three postulates. On the basis of this definition, which is simply an explicit statement of the fact that the angle output is independent of the amplitude level of the received signal and that it has odd symmetry about the boresight, all known types of monopulse may be derived as special cases of a completely general theory. The special theory associated with these cases does even more than provide a consistent explanation of the origin of the known types of monopulse and of their interrelationships; it also shows that they form a mutually exclusive set, i.e., that these *and no more* are possible within the postulated structure of the theory. Perhaps of even greater importance to the future development of monopulse is the discovery of a general theory. Although playing only a minor role in this monograph, it may prove to be of more far-reaching significance than that which led to its discovery, namely, an attempt to unify the known types of monopulse into a simple, consistent theory.

The principal result of the general theory is that the operation of a monopulse system can be described analytically by a sequence of mapping transformations of a complex variable from one complex plane to another. Two transformations are common to all monopulse systems, namely, mapping of a fixed function of angle of arrival onto the angle-sensing plane, and mapping of the sensing ratio onto the angle-detection plane. The first is controlled by the angle sensor, the second by the angle detector. A third transformation may be performed in between these two basic transformations in order to convert the sensing ratio formed by the angle sensor into whatever form is desired for angle detection.

In the general theory the sensing-ratio and the angle-detection functions may be any arbitrary complex functions satisfying the monopulse postulates. The special theory is restricted specifically to those particular cases of the general theory that involve either pure amplitude or pure phase sensing. Sensing in the special theory is always described by one of three functions, amplitude $\rho(u)$, phase $\phi(u)$, or sum-and-difference $\Delta(u)/\Sigma(u)$. The sum-and-difference sensing function is related to both the amplitude- and phase-sensing functions, a relationship which implies an equivalence between the latter two. Thus there are just two independent, although equivalent, forms of special sensing. When limited to these particular forms there are three, and only three, distinct classes of angle detection possible. Either pure amplitude or pure phase sensing may be used with each of the three classes of angle detection, but since amplitude and phase sensing are equivalent there are just three distinct classes of special monopulse possible.

3. Monopulse systems

A GENERAL theory has been derived from three fundamental postulates defining the concept of monopulse. From this the basic properties of a special theory were outlined, the special theory being limited specifically to the cases of pure amplitude or pure phase sensing in any single plane. These include all known forms of monopulse. Thus the foundation for a unified theory of the monopulse concept has been laid. To build upon this foundation, and to further the development of the special theory in particular, we shall now examine some of the techniques by which the various forms of monopulse can be realized physically. The present treatment is not intended to be exhaustive but rather to illustrate, by way of specific examples, ways in which each form of monopulse can be realized. The examples considered here will be limited to two-dimensional operation in a single principal plane of incidence. Dual-plane systems for complete three-dimensional operation will be treated in the next chapter.

3.1 Functional system components

The basic components of a monopulse system may be blocked out as in Fig. 3.1 according to the sequence of functions performed (Fig. 2.1). The angle of arrival is sensed by the monopulse antenna, which is the primary, and sometimes the only, element of the angle sensor. The angle information appears in the form of two functions whose ratio, the sensing ratio, is formed later. The two functions must either be mirror images of each other (multiplicative sensing) or one must have odd and the other even symmetry (additive sensing) about the boresight axis. After sensing the angle of arrival, the sensing ratio may be converted into any other form desired by performing a bilinear transformation (which may be just the identity transformation) and then detected to obtain the angle-output function. Ultimately the two functions sensed will appear only as a ratio; hence their conversion and detection are equivalent to conversion and detection of their ratio. The purpose of ratio conversion may either be to convert the angle information from amplitude to phase, or vice versa, in order to match the angle sensor to the angle detector, or to control the range of angle-scanning or the boresight

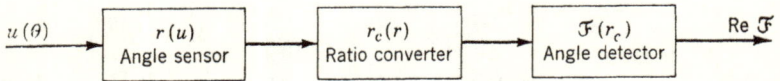

$u(\theta)$ → | $r(u)$
 Angle sensor | → $r_c(r)$ | $r_c(r)$
 Ratio converter | → $\mathcal{F}(r_c)$ | $\mathcal{F}(r_c)$
 Angle detector | → Re $\mathcal{F}$

Fig. 3.1 Functional block diagram of a monopulse system.

sensitivity, or both. It is usually performed at the carrier frequency using passive components because of the relative simplicity and stability obtainable. When the system is used in a radar, duplexing can be introduced either in the angle sensor or in the ratio converter. All the rest of the system following the ratio converter will be lumped together under the term of angle detector. The angle detector usually contains all of the active components in the system, including the amplifiers and comparison circuitry that form the ratio and the angle-detection function and that produce the real part of the angle-detection function at the output.

3.2   Angle sensors

The source of raw data on the angle of arrival of an incident wave is the pair of signals extracted from that wave by the sensing antenna. These may be in the form of complex functions of angle of arrival, as in the general theory of monopulse, or they may be simple functions that differ only in amplitude or in phase, but not both, as in the special theory. The field of complex sensing remains virtually untouched at the present time, all available effort having been directed toward pure amplitude and phase sensing. Therefore two of the more common types of antennas in use for simple sensing, the reflector and the lens illuminated by directional point sources, will be described briefly.

Typical monopulse antennas

A parabolic reflector excited symmetrically by a primary source at the focal point will produce a symmetrical pattern with a spherical phase front. Two such reflectors (Fig. 3.2a) constitute an interferometer with phase centers spaced a distance s apart.

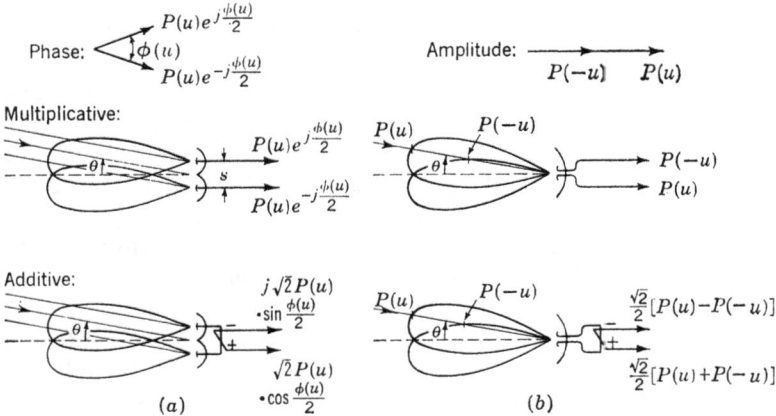

Fig. 3.2 Monopulse reflector-type antennas for (a) phase and (b) amplitude sensing of the multiplicative and additive sensing functions.

Their amplitude patterns will be identical but their phase patterns will differ by

$$\phi(u) = \frac{2\pi}{\lambda} s \sin \theta$$

$$= \frac{2s}{d} u. \tag{3.1}$$

The interferometer-type antenna can be used for pure phase sensing. The signals received can be used directly for multiplicative phase sensing, or for additive phase sensing by forming their sum and difference in a hybrid T as shown. An example of an interferometer with parabolic reflectors was shown in Fig. 1.7.

If the primary source illuminating the parabolic reflector is displaced laterally from the focal point f by a small distance Δx, the phase center will remain close to its original position but the amplitude pattern will be squinted off of the boresight axis at an angle of approximately $\Delta x/f$, the angle of specular reflection from a flat mirror (Fig. 3.3). A pair of feeds displaced symmetrically from the focal point will then produce symmetrically overlapping amplitude patterns (Fig. 3.2b) whose ratio indicates angle of arrival. The two received signals can be used directly for multiplicative amplitude sensing, or by forming their sum and difference for additive amplitude sensing. Amplitude sensing by off-focus feeds has the distinct advantage of simplicity but is never quite pure because of the fact that the phase centers can never quite be brought into coincidence. Examples of this type of sensor are shown in Figs. 1.8 and 1.9.

For each reflector-type of monopulse antenna there corresponds an analogous lens-type. The lens is somewhat more versatile than the reflector in that its aperture distribution, and hence its pattern, is usually capable of finer control. The phase distribution may be controlled by varying the index of refraction as well as the shape of the lens, and fine control of

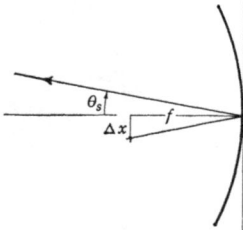

Fig. 3.3 Squint angle produced by specular reflection in a parabolic reflector.

the amplitude distribution may be obtained by introducing attenuation into the refracting medium across the lens. A phase-sensing interferometer may be constructed of a pair of lenses with displaced feeds located at their respective focal points, and amplitude-sensing squinted beams may be generated by a single lens with off-focus feeds (Fig. 3.4). Although the lens- and reflector-type antennas are similar in effect, there are significant practical

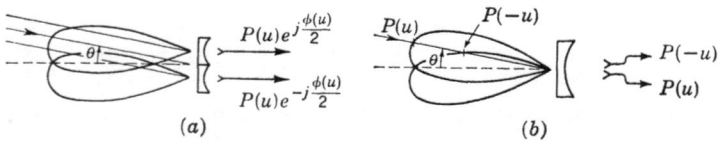

Fig. 3.4 Monopulse lens-type antennas for (a) phase and (b) amplitude sensing.

differences between them. Usually the aperture of a reflector is partially blocked by the feed, while the aperture of a lens is not. With the feed and waveguide out of the primary field there is no distortion of the aperture distribution and no limitations on the structural support of the feed. On the other hand, the reflector-type antenna may be fitted into a smaller volume, is lighter, and is usually easier to fabricate. Both have their advantages and disadvantages; in any particular application

the choice will depend ultimately upon which of these characteristics can be compromised most readily.

3.3 Ratio converters

Any of the four special sensing ratios may be converted into any
of the others, but all but two of the conversions reduce to degenerate forms of the original four (the two additive ratios themselves
have been interpreted as conversions of the two multiplicative
ratios). The two nondegenerate conversions are those between
multiplicative amplitude and phase. They can be realized
physically by a short slot coupler* (Fig. 3.5a). If the ratio of
the two input signals is r, then the ratio of the output signals
from the coupler is the bilinear transformation $(jr + 1)/(r + j)$.

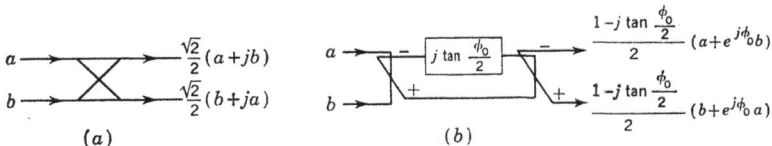

Fig. 3.5 Circuits for converting the sensing function: (a) the short slot
coupler for the special case of conversion between amplitude and phase, and
(b) a general conversion network.

The net effect of the short slot coupler is to convert the amplitude-
sensing function into its equivalent phase-sensing function, and
vice versa, without changing its realizable angular range or boresight sensitivity.

In its broadest sense the purpose of ratio conversion may be
more than transformation of one sensing function into another.
It may also be used to alter the realizable angular range or boresight sensitivity and to provide a means for duplexing (Sec. 3.6).
A network for realizing all three simultaneously is shown in Fig.
3.5b. The difference of the two received voltages is advanced 90°
in phase and attenuated (or amplified) by the factor $\tan (\phi_0/2)$,
after which a new sum and difference is formed. When the ratio
of the input signals is the amplitude-sensing function $\rho(u)$, the
ratio of the output signals can be shown to be

$$\exp j2 \tan^{-1} \left[\frac{\rho(u) - 1}{\rho(u) + 1} \tan \frac{\phi_0}{2} \right].$$

---

* H. J. Riblet, The Short Slot Hybrid Junction, *Proc. IRE*, vol. 40, pp.
180–184, February, 1952.

Thus the amplitude-sensing function is converted into a phase-sensing function [Eq. (2.17)] over an angular range determined by ϕ_0, or by its equivalent ρ_0. Conversely, the same coupler will convert the phase-sensing function into an amplitude-sensing function over an angular range also determined by ϕ_0.

Examples of general conversion networks for each of the six nondegenerate forms of monopulse are illustrated in Fig. 3.6. The two class I conversions degenerate into a common form; hence these two sensors convert into each other. The class IIP

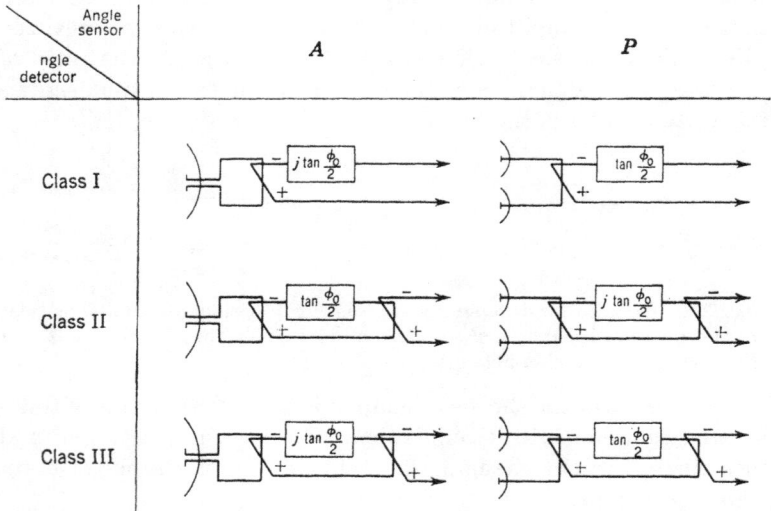

Fig. 3.6 Angle sensors with general conversion circuits for each of the six nondegenerate forms of monopulse.

and IIIA conversion networks shown are just those described previously in Fig. 3.5b. When converting from amplitude to amplitude (class IIA) or from phase to phase (class IIIP), however, double conversion is necessary in order to return to the original sensing function. Double conversion can be performed by cascading two networks like the one shown in Fig. 3.5b. When the two are cascaded, they reduce to the double-conversion network shown.

For the important case of conversion of the sensing function without change in angular range, i.e., $\phi_0 = \pi/2$, the two networks in Fig. 3.5 are essentially alike electrically. The particular choice of network in this case will depend on the application.

There is a very real structural advantage of the short slot coupler, since it can consist simply of one or more slots cut in the wall between the two wave guide feeds (either the narrow wall* or the broad wall†).   On the other hand, formation of the sum-and-difference pattern functions has some definite advantages in duplexing and mechanical scanning, as discussed in Sec. 3.6.

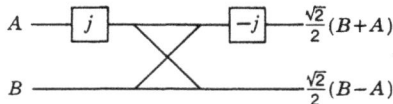

Fig. 3.7   Short slot network equivalent to the hybrid T.

Both advantages can be obtained by forming the sum and difference with a short slot coupler (Fig. 3.7) instead of a hybrid T.

3.4   Angle detectors

Only three classes of angle detection are admitted by the special theory.   There is an unlimited number of specific forms of angle detection possible for each class, however, the only restriction being that the angle-output function (real part of the angle-detection function) be odd and that it and its derivatives be continuous.   The present section will describe some of the forms of angle detection that are used in practice.

Typical circuits

Circuits for each of the three classes of angle detection illustrated in Sec. 2.5 are blocked out schematically in Fig. 3.8.   In each case the two r-f signals from the ratio converter are heterodyned with a local oscillator (LO) to produce an intermediate frequency suitable for amplification.   By using a common LO for the two channels, symmetry of the two signals is preserved and, of importance in classes I and III, phase coherence between channels is maintained.   After mixing, the two signals are amplified and compared.

The angle-detection function for the class I angle detector illustrated is simply the sensing ratio itself in the case of amplitude

* H. J. Riblet, Waveguide Hybrid, U.S. Pat. 2,739,288, filed Mar. 17, 1950, issued Mar. 20, 1956.

† H. J. Riblet, Hybrid Directional Coupler, U.S. Pat. 2,709,241, filed Feb. 28, 1950, issued May 24, 1955.

sensing, and the sensing ratio rotated 90° in the case of phase sensing (Fig. 2.9). Any function with an odd real part could have been chosen; the one actually chosen is the simplest. Since the ratio lies on the real (or imaginary) axis, the angle output of the system is just the angle-detection function $\Delta(u)/\Sigma(u)$.

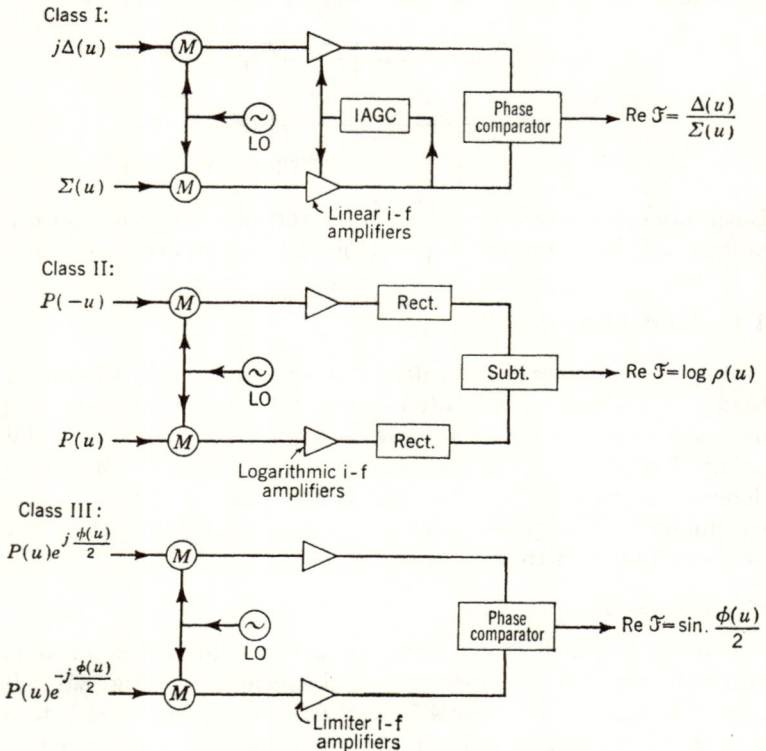

Fig. 3.8 Examples of angle-detection circuits.

The class II angle detector illustrated is described analytically by the logarithmic function, although any other function whose even derivatives at $r_m = 1$ have real parts given by Eq. (A.8) is acceptable. The logarithmic function is obviously acceptable since $\log r_m(u)$ vanishes on the boresight $r_m(0) = 1$, and has odd symmetry about the boresight. The angle-detection function is $\log \rho(u)$, which is real; hence the angle output is again the angle-detection function itself.

The class III angle detector illustrated may be described by the function $-jr_m^{1/2}(u)$ over a portion of the unit circle. Its real part is Re $[-je^{j\phi(u)/2}] = \sin [\phi(u)/2]$. Since the angle-detection function and all of its derivatives are purely imaginary on the boresight, its real part and the real part of all of its derivatives vanish there and therefore satisfy the condition of continuity.

Normalization

The ratio is formed by the normalization property of the amplifiers. In the class I angle detector illustrated this property is introduced in the form of instantaneous automatic gain control (IAGC). An IAGC voltage is developed in the sum channel and used to control the gain of both channels. The result, in effect, is to normalize both the sum- and the difference-signal amplitudes with respect to the amplitude of the sum signal. The output of the sum channel then remains constant in time, ideally, and the output of the difference channel becomes the ratio $\Delta(u)/\Sigma(u)$. In the class II angle detector normalization is obtained in quite a different way. Here the ratio is formed, in effect, by subtracting magnitudes of logarithms to produce the logarithm of the ratio. On the boresight the ratio is unity and hence the logarithmic angle output is zero, as required. Since the logarithmic outputs from the amplifiers indicate only relative amplitudes, they are inherently independent of any absolute scale. The phase relation between the two input signals does not affect the ratio since the signals are demodulated before taking their ratio (difference of logarithmic amplitudes). Class III normalization is different again. All amplitude variations must be removed from both channels; this is effected by the use of limiter-type amplifiers that give constant and equal output amplitudes regardless of variations in amplitude of the signals received. An alternative way to achieve the same effect in principle is to use IAGC on both amplifiers, either a common IAGC voltage derived from one amplifier and applied to both, as in the illustration of class I, or independently derived IAGC voltages each applied to their own amplifier. Both ways, limiting and independent IAGC in each channel, have an important advantage over common IAGC in that the two signals to the angle detector need not be of equal amplitude. An example of a class III angle detector with independent IAGC has been described by Sommer[5] for use in a missile tracking seeker. Thus in the case of either classes II or III simple sensing is obtained, in effect, independently of the nonsensing patterns of the antenna.

Formation of the angle-detection function

Just as the ratio in the three classes of angle detection is formed in quite different ways, so is the angle-detection function itself. In the case of class II it is formed by the logarithmic character-istic of the amplifiers, the same characteristic that, in effect, forms the ratio. In the other two classes it is formed solely by the action of the phase comparator. For the angle detectors illustrated the comparator may be either the conventional phase comparator* or the extended-range phase comparator invented by Kirkpatrick.† The conventional phase comparator is shown

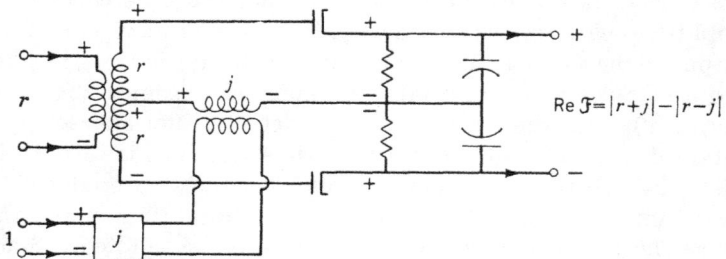

Fig. 3.9 The conventional phase comparator.

in Fig. 3.9 (a 90° phase shift has been included in order for the output to have odd symmetry). When complex voltages are applied at the two pairs of input terminals, their sum and their difference are formed by the hybrid network and then rectified by linear detectors. The difference of magnitudes of the sum and the difference then appears as a real function at the output (in accordance with the third postulate). The angle output is proportional to

$$\text{Re } \mathfrak{F}(r) = |r + j| - |r - j|. \tag{3.2}$$

It satisfies the monopulse postulates when r is any arbitrary additive ratio, real or complex, and hence can be used for complex additive angle detection as well as for the cases of simple angle detection considered here. This is not so when r is any arbitrary

---

* R. H. Dishington, Diode Phase-discriminators, *Proc. IRE*, vol. 37, pp. 1401–1404, December, 1949.

† G. M. Kirkpatrick, Extended-range Phase Comparator, U.S. Pat. 2,751,555, filed Oct. 3, 1951, issued June 19, 1956.

multiplicative ratio. For an arbitrary multiplicative ratio $\rho(u)e^{i\phi(u)}$, Eq. (3.2) may be expressed as the real part of

$$\mathfrak{F}(r_m) = -j \sqrt{\rho^2(u) + 1} \exp\left(j\left\{\frac{\pi}{4} - \frac{1}{2} \cos^{-1}\left[\frac{2\rho(u)}{\rho^2(u) + 1} \sin \phi(u)\right]\right\}\right).$$

It can be seen here that the real part of $\mathfrak{F}$ at r_m is not the negative of its real part at $1/r_m$ except on the unit circle $|r_m| = 1$, i.e., pure phase sensing. In the case of simple angle detection, however, the sensing ratio for class I is $r_a(u) = j\,\Delta(u)/\Sigma(u)$ and the angle output is proportional to

$$\text{Re } \mathfrak{F} = \frac{\Delta(u)}{\Sigma(u)}, \qquad -1 \leq \frac{\Delta}{\Sigma} \leq 1.$$

For class III the sensing ratio is $r_m(u) = e^{i\phi(u)}$, and the angle output is proportional to

$$\text{Re } \mathfrak{F} = \sin \frac{\phi(u)}{2}, \qquad -\frac{\pi}{2} \leq \phi \leq \frac{\pi}{2}.$$

Angular-scanning range

The range of angle scanning with the conventional phase comparator is limited to that corresponding to a range of phase of 180°, but the maximum unambiguous range of the angle sensor itself is frequently greater than this. It has already been seen (Sec. 2.3) that the range of phase can be extended to a maximum of 360° by r-f conversion of the sensing ratio, but it is accompanied by vanishing sensitivity. Since it involves attenuation of the r-f difference signal, it will be accompanied by a loss in signal-to-noise ratio in the difference channel and hence a loss in radar range. Another method of extending the angular range, this time in the comparator itself, was introduced with the Kirkpatrick comparator. As with ratio conversion, the Kirkpatrick comparator achieves extended range only at the expense of boresight sensitivity, but it does so with little or no loss in signal-to-noise ratio because it operates at the point of highest amplitude level in the system. The essence of the Kirkpatrick comparator is illustrated in Fig. 3.10. Its detection portion is the same as in the conventional phase comparator, consisting of linear rectifiers and a differencing network. The essential difference between the Kirkpatrick and the conventional phase comparators lies in the way the signals are combined before detection. In both cases the angle output is the difference between magnitudes of the complex voltages appearing at the detector diodes:

$$\text{Re } \mathfrak{F} = |v_1| - |v_2|.$$

In the conventional phase comparator the voltages v_1 and v_2 relative to the common terminal are seen to be just the sum and the difference of the input voltages. They are not seen as easily in the Kirkpatrick comparator, but can be calculated by superposition of the components generated by each of the input voltages. The components v_1' and v_2' generated by the input voltage

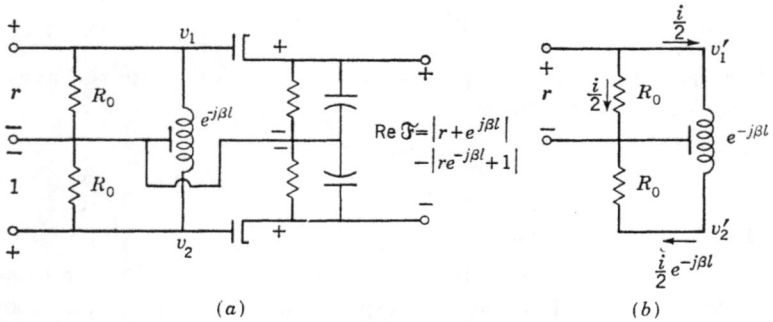

Fig. 3.10 The Kirkpatrick comparator: (a) simplified circuit of the complete comparator, and (b) portion driven by the source r.

r, for example, involve only those parts of the circuit in Fig. 3.10b. Current i, generated by r, splits at the junction of two parallel impedances. One is made up of a transmission line of delay βl terminated in its characteristic impedance R_0, the other is R_0 itself. Since the input impedance of a transmission line terminated in its characteristic impedance is just its characteristic impedance, the current i splits equally into the two resistances. The two components generated by r are then

$$v_1' = r$$

and
$$v_2' = re^{-j\beta l}.$$

Similarly, but in reverse order, for the components generated by the other input voltage. By superposition, then,

$$\text{Re } \mathfrak{F} = |r + e^{-j\beta l}| - |re^{-j\beta l} + 1|. \qquad (3.3)$$

In the case of the class I angle detector $r = j[\Delta(u)/\Sigma(u)]$; hence the angle output is proportional to

$$\text{Re } \mathfrak{F} = \left| j\frac{\Delta(u)}{\Sigma(u)} + e^{-j\beta l} \right| - \left| j\frac{\Delta(u)}{\Sigma(u)} + e^{j\beta l} \right|. \qquad (3.4)$$

In the case of the class III angle detector $r = e^{j\phi(u)}$; hence the angle output is proportional to

$$\text{Re } \mathfrak{F} = \left| e^{j\phi(u)} + e^{-j\beta l} \right| - \left| e^{j\phi(u)} + e^{j\beta l} \right|$$

$$= 4 \sin\frac{\beta l}{2} \sin\frac{\phi(u)}{2}, \qquad -(\pi - \beta l) \le \phi \le (\pi - \beta l). \quad (3.5)$$

When $\beta l = \pi/2$, both cases reduce identically to those of the conventional phase comparator. But for $\beta l < \pi/2$ the angular

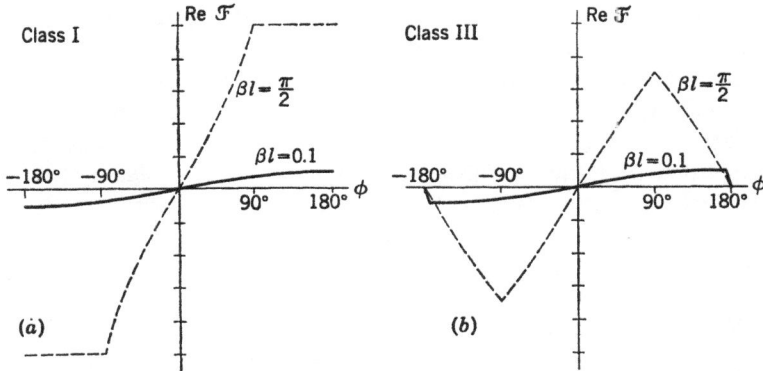

Fig. 3.11 Angle-output functions from the Kirkpatrick comparator.

range is extended to a value approaching the maximum of 360° as βl approaches zero. In the case of the class III angle detector the angle-output function is just an extension of the original $\sin[\phi(u)/2]$ function, but is attenuated by the factor $\sin[\beta l/2]$ which falls to zero at the maximum angular range. In this case the angle output is proportional to the real part of

$$\mathfrak{F}(r_m) = -j4 \sin\frac{\beta l}{2} r_m^{1/2}(u). \qquad (3.6)$$

The characteristics of the angle output from a conventional phase comparator and from the Kirkpatrick extended-range phase comparator may be observed and compared in Fig. 3.11 for both classes I and III. The unambiguous angular range is limited by the conventional phase comparator to a range of phase of $-90° \leq \phi \leq 90°$ for both classes. So is the Kirk-patrick comparator if $\beta l = \pi/2$, for all electrical characteristics of the two are then identical. When $\beta l < \pi/2$, however, the angular range is extended beyond that of the conventional com-parator to the full 360° in the case of class I, and to $-(\pi - \beta l) \leq \phi \leq (\pi - \beta l)$ in the case of class III. The example of angular-range extension is for $\beta l = 0.1$, a rather small value resulting in a large extension in range and a correspondingly large decrease in sensitivity. The class III angular range has been extended to 348°, but the boresight sensitivity for both classes has been reduced by a factor of ten. A rather interesting characteristic of the Kirkpatrick comparator for small βl is that its angle output for both classes is nearly identical, the two functions approaching $2\beta l \sin [\phi(u)/2]$ over nearly the full 360°. The theoretical maxi-mum range can be approached as closely as desired by the Kirk-patrick comparator, and although it is obtained only by a cor-responding loss in sensitivity, this loss occurs in the system at a point where the signal-to-noise ratio is its greatest. Thus it is concluded that the unambiguous angular range for classes I and III need be limited only to that of the monopulse antenna.

3.5 Monopulse systems

Now that some of the techniques in common use for sensing, conversion, and detection of monopulse angle information have been seen, it will be instructive to assemble and compare exam-ples of all of the various monopulse systems permitted by the special theory. A complete set of examples is summarized in Fig. 3.12. They are classified according to the three forms of angle information to the angle detector, namely, class I associated with the additive sensing function $\Delta(u)/\Sigma(u)$, and classes II and III associated with the multiplicative sensing functions $\rho(u)$ and $\phi(u)$, respectively, or their equivalents. The three classes are then divided according to the type of sensing, either ampli-tude or phase, giving the six distinct forms of monopulse. For each sensing function one type of sensing is described by the actual sensing function and the other by its equivalent. In the examples shown, each of the actual sensing functions is con-verted into its equivalent without change in angular range or boresight sensitivity.

Equivalence between angle sensors is indicated in the figure by two-way arrows. In the case of the two multiplicative classes II and III there is just one amplitude and one phase sensor that are equivalent to each other. In the case of the two additive classes, however, all four are equivalent. The ratio of outputs from all four is of the same form, namely, $j\Delta(u)/\Sigma(u)$. This illustrates the degeneracy pointed out in Sec. 2.3 between two of the four forms of the additive ratio. Only two of the four are distinct,

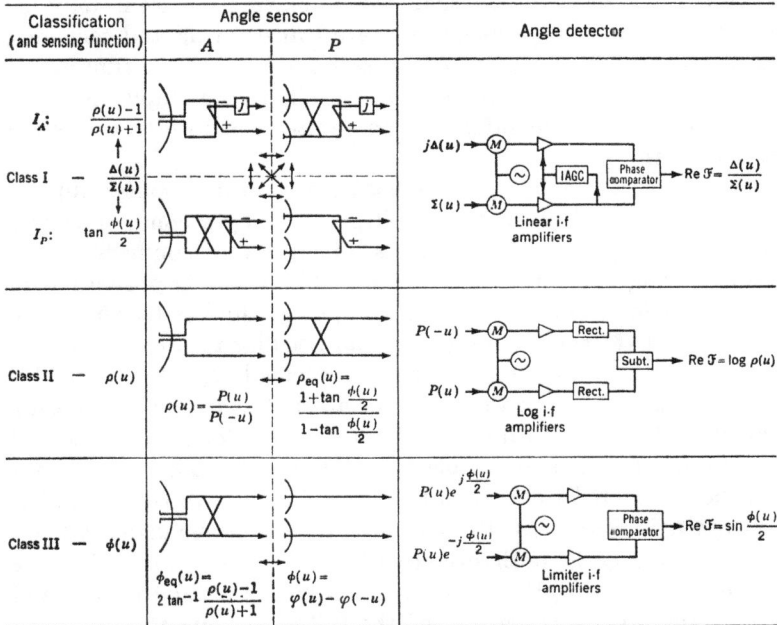

Classification (and sensing function)	Angle sensor		Angle detector
	A	P	
I_A: $\frac{\rho(u)-1}{\rho(u)+1}$ Class I $\frac{\Delta(u)}{\Sigma(u)}$ I_P: $\tan\frac{\phi(u)}{2}$			$j\Delta(u)$, $\Sigma(u)$, Linear i-f amplifiers, IAGC, Phase comparator, $\mathrm{Re}\,\mathfrak{F}=\frac{\Delta(u)}{\Sigma(u)}$
Class II — $\rho(u)$	$\rho(u)=\frac{P(u)}{P(-u)}$	$\rho_{eq}(u)=\frac{1+\tan\frac{\phi(u)}{2}}{1-\tan\frac{\phi(u)}{2}}$	$P(-u)$, $P(u)$, Log i-f amplifiers, Rect., Subt., $\mathrm{Re}\,\mathfrak{F}=\log\rho(u)$
Class III — $\phi(u)$	$\phi_{eq}(u)=2\tan^{-1}\frac{\rho(u)-1}{\rho(u)+1}$	$\phi(u)=\varphi(u)-\varphi(-u)$	$P(u)e^{j\frac{\phi(u)}{2}}$, $P(u)e^{-j\frac{\phi(u)}{2}}$, Limiter i-f amplifiers, Phase comparator, $\mathrm{Re}\,\mathfrak{F}=\sin\frac{\phi(u)}{2}$

Fig. 3.12 Examples of the six distinct forms of monopulse.

one giving the ratio $j[\rho(u) - 1]/[\rho(u) + 1]$ and the other the ratio $j\tan[\phi(u)/2]$.

These examples of the six distinct forms of monopulse admitted by the special theory are by no means the only ones possible. In fact, an unlimited number of sensing functions, ratio conversions, and angle-detection functions is possible.

3.6 Duplexing and scanning

Since the concept of monopulse is limited strictly to reception, little mention has been made as yet of the problems of introducing a transmitter into the system when it is to be used as part of a radar. Generally speaking, the transmitter portion of a monopulse radar is much the same as in any conventional pulse radar. Two important differences do exist, however, in the problems associated with duplexing and mechanical scanning. Monopulse duplexing is complicated by the necessity for multiple antenna feeds and receiving channels, in contrast to the usual single-feed and receiving channel of conventional radar. It was seen in Sec. 1.4 that the duplexing problem may be avoided altogether by using separate transmitting and receiving antennas,[1,6] but in many cases this is far from the optimum use of the antenna aperture available. Therefore duplexing is a necessity if the same antenna is to be used for both transmission and reception.

Two general problems are associated with monopulse duplexing. One is that of exciting multiple feeds by a common source in such a way that the source is effectively disconnected from them during the interpulse period. The other is protection of the receivers from the peak pulse power during transmission. The first of these is the only problem peculiar to monopulse. If the sum and difference of the signals received are formed directly at the antenna, which was done for all six forms of angle sensing shown in Fig. 3.6, both of the duplexing problems can be solved by use of conventional duplexing techniques. The transmitter can be connected to the sum line to excite the two feeds equally, as shown in Fig. 3.13, thereby producing a transmitted pattern with symmetry in both principal planes. On transmission the pulse power opens the anti-transmit-receive (ATR) switch and shorts the TR switches to feed the full power into the sum arm, which then splits equally at the two antenna feeds. On reception the ATR switch disconnects the transmitter from the sum line while the TR switches open to pass the sum and difference of the received signals unimpeded to the angle detector, either directly or through a suitable ratio-conversion network.

The other feed problem peculiar to monopulse occurs when the antenna must rotate for mechanical scanning. If the rest of the receiving system can be made to rotate with the antenna there is no problem; a single conventional rotary joint for the transmitter alone is sufficient. In the event that the rest of the system must remain fixed, however, as it frequently must in practice, it becomes necessary to use a dual-channel rotary joint. This complicates the feed system still further and raises the question of where it should be placed in the circuit in order that it will introduce the least error in the angle measurement. If it were placed directly in the antenna feed lines, any amplitude or phase error introduced by the changing geometry of the rotary joint would be reflected as an error in the angle measurement. If, on the other hand, it were placed in the sum-and-difference lines, the angle-output function would still be affected but its boresight indication would remain unchanged. Since preservation of the boresight is of utmost importance in most applications, this is an important virtue. A more detailed discussion of the dual-channel rotary-joint problem has been given by Raabe.* He describes a method for designing dual-channel joints by coupling two rectangular

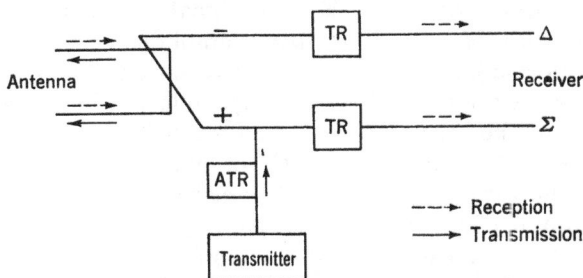

Fig. 3.13 A sum-and-difference monopulse radar duplexer.

wave guide channels into a single circular wave guide capable of supporting two circularly symmetrical modes. Another type of dual-channel rotary joint was developed earlier by Farr,† based on the use of oppositely rotating circularly polarized waves in a single circular wave guide.

In summarizing the problems of duplexing and scanning one point stands out: a common solution to both lies in the use of a sum-and-difference network at the antenna.

3.7   Pseudomonopulse

The entire theory of monopulse as developed in this monograph is an idealization that represents an ultimate objective rather

* H. P. Raabe, A Rotary Joint for Two Microwave Transmission Channels of the Same Frequency Band, *IRE Trans. PGMTT*-3, pp. 30–41, July, 1955.

† H. K. Farr, Two Channel Rotary Joint, U.S. Pat. 2,713,151, filed Mar. 29, 1946, issued July 12, 1955.

than a physical reality.   In actual practice this idealization can
be realized only approximately by any physical system.   Within
the limitations of physical realization, then, any simultaneous-
lobing pulse-receiving system designed to approach this ideal
may be considered as an approximation to true monopulse.
There are other systems, however, that are frequently referred
to as being monopulse, and that have at least some of the characteristics of monopulse as defined here, but that are based on a
fundamentally different design philosophy.   Such systems will be
termed "pseudomonopulse."

The most common difference between pseudomonopulse and
true monopulse as defined here is violation of the first postulate.
This may be a result either of only partially effective, or of totally
lacking, normalization of the two signals.   Identical dynamic
characteristics of the normalizing amplifiers are essential to
effective normalization.   For the particular amplifiers illustrated
in Sec. 3.5 they must be linear in amplitude and have instantaneous AGC in the case of class I, or they must be logarithmic in
amplitude in the case of class II, or linear in phase and have
instantaneous limiting action in the case of class III.

An example of a simultaneous-lobing system that is frequently
considered as a form of pseudomonopulse (with totally lacking
normalization) is a beam-narrowing system that appeared early
in the history of radar.[8]   By 1940 the British had built and tested
a 200-Mc search radar based on a "split-beam" technique.
Sum-and-difference amplitude-sensing patterns from a five-
element array of dipoles were obtained (Fig. 3.14a), and after
amplification in separate linear amplifiers were rectified and their
magnitudes subtracted (Fig. 3.14b).   The difference between the
sum and the difference amplitude patterns caused partial can-
cellation of return from targets near the edges of the beam and
gave an effective beamwidth that was narrower than that of the
sum pattern alone.   By manual adjustment of the difference
amplifier gain it was possible to control the effective beamwidth
of the antenna.   At zero gain the beamwidth was that of the sum
pattern alone, while at high gain the beamwidth was reduced
considerably but the side-lobe level increased.   Since the side
lobes of the transmitting pattern appear at different angles from
those of the effective receiving pattern, the side-lobe level of the
over-all two-way pattern is suppressed considerably (Fig. 3.14c).
This is an example of simultaneous lobing with symmetry about

the boresight but with complete absence of normalization, and necessarily so, since normalization would destroy the very information sought by a search radar.

Examples of pseudomonopulse in which normalization is only partially effective are offered by some boresight-tracking radar systems. A class I system with slow rather than instantaneous AGC would still give an accurate indication of the boresight by the position of the *crossover*, or null, in the angle-output function, but the slope (boresight sensitivity) would vary with the amplitude level of the return wave. If the resulting angle output is to supply an error signal to a servo having either an extremely short

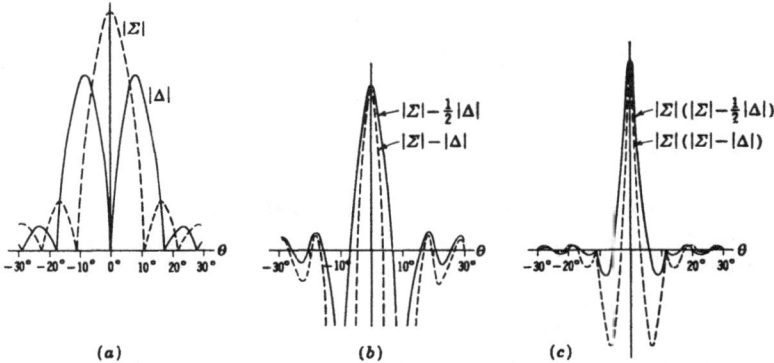

Fig. 3.14 British search-radar beam-narrowing system: (c) rectified sum-and-difference antenna patterns, (b) narrowing of the effective receiving pattern, for full-difference gain and for half-difference gain, and (c) the combined transmit-receive pattern. (*After Taylor and Wescott.*)

time constant (such as an electron beam) or a time constant long relative to the interpulse period, then it may be unimportant whether or not the system operates as true monopulse. If, on the other hand, the servo time constant is comparable to the interpulse period, then the changing sensitivity may result in angular jitter about the boresight. A specific example of a class I radar system with one form of slow AGC is a system proposed by Smith and Brockner.[7] Here the AGC is purposely delayed by a full pulse period, the return-signal strength from one pulse controlling the gain of the amplifiers on reception of the next pulse.

A not uncommon form of pseudomonopulse is one possessing both normalization and symmetry, but symmetry that is even

rather than odd.   Such a system would have a most serious lack
of sense indication.   Nevertheless it does have important applications, one of which has been described by Busignies* for improv-
ing the angular accuracy of a search radar.   Here the comparison
of signals received is limited solely to an indication of coincidence
between amplitudes of two signals.   As the radar boresight is
made to scan past a target, the azimuthal direction of that target
is indicated quite accurately on a pulse-position-indicator (PPI)
by intensifying the spot in the direction at which coincidence
occurs.

Although not as common as violation of the first postulate,
some systems are not true monopulse because of a lack of symmetry about the boresight.   An important example of this form
of pseudomonopulse is electronic boresight scanning, described in
the next section.

3.8   Electronic boresight scanning

In many applications it is advantageous, and sometimes
mandatory, that scanning be performed electronically by a fixed
antenna rather than by the usual brute-force method of rotating
the antenna mechanically.   Electronic scanning is certainly a
necessity whenever the required scanning rate exceeds the maximum attainable by mechanical means, and is highly desirable,
if not absolutely necessary, for any precision tracking system in
which the problems associated with radomes and mechanical
scanners are especially severe.   The direct approach to electronic
scanning is to produce a traveling-wave aperture function whose
phase velocity can be varied.   The beam itself then scans in
response to the changing phase velocity.   This technique is
satisfactory in many applications but is limited in accuracy to
that of any conventional scanning system.   Another approach is
based on the use of monopulse.   Monopulse is inherently an
electronic-scanning concept and so is ideally suited to such
applications.   Angle-scanning monopulse is limited in stability
by its dependence on amplifier characteristics, however, and
boresighting monopulse, per se, is incapable of electronic scanning.
It would be highly desirable to retain the inherent stability of
the class I boresight in a monopulse system capable of electronic

---

* H. G. Busignies, Signal Comparison System, U.S. Pat. 2,509,207, filed
Apr. 26, 1944, issued May 30, 1950.

scanning. This can be done by scanning the *indicated boresight* electronically, as will be shown here.

The boresight direction is indicated by the value of u at which $r_m(u) = 1$. The indicated boresight direction can be either the actual boresight direction $u = u_b$ of the antenna (it has been implied so far that u_b is zero, i.e., that the boresight is broadside to the aperture, but it will be shown in Sec. 5.4 that it may have any given value), or it can be any other selected direction by transforming the sensing ratio in that direction to unity. The value of u associated with the point $r_{ms}(u) = 1$ in the plane of the transformed sensing ratio then represents the boresight direction that will be indicated by a null in the angle output. This is the theoretical basis of electronic boresight scanning.· When restricted to pure amplitude or pure phase sensing, the scanning transformation corresponds, respectively, to a shift along the real axis or around the unit circle. Thus electronic boresight scanning can be realized physically by attenuation (amplification) or phase shift, respectively, of one signal relative to the other:

$$r_{ms}(u) = \begin{cases} e^{\alpha}\rho(u) & \text{amplitude sensing,} \\ e^{i\beta}e^{i\phi(u)} & \text{phase sensing.} \end{cases} \qquad (3.7)$$

This is illustrated in Fig. 3.15. The actual boresight appears at e^{α} (or $e^{i\beta}$), but the boresight as indicated by a null in the angle output still remains fixed at unity. By varying α (or β), the indicated boresight direction can then be controlled at will.

For each value of scanning attenuation α there is an equivalent value of scanning phase shift β that will indicate the same boresight direction. Any amplitude function (including the sum-and-difference function) may be converted to an equivalent phase function and its boresight shifted by shifting phase by an amount β before converting back to amplitude as in Fig. 3.15c, and conversely for phase functions as in Fig. 3.15d. From Eq. (3.7) the shift in indicated boresight resulting from a shift in phase β is

$$\phi(u_{ib}) = -\beta, \qquad (3.8a)$$

while that resulting from an attenuation α is

$$\rho(u_{ib}) = e^{-\alpha}. \qquad (3.8b)$$

In view of the equivalence [Eq. (2.13)] between the amplitude- and phase-sensing functions, there exists a corresponding equivalence between the scanning attenuation and scanning phase shift:

$$\tan \frac{\beta}{2} = \tanh \frac{\alpha}{2}. \tag{3.9}$$

Electronic boresight scanning is applicable to any of the three classes of angle detection. The scanners in Figs. 3.15a

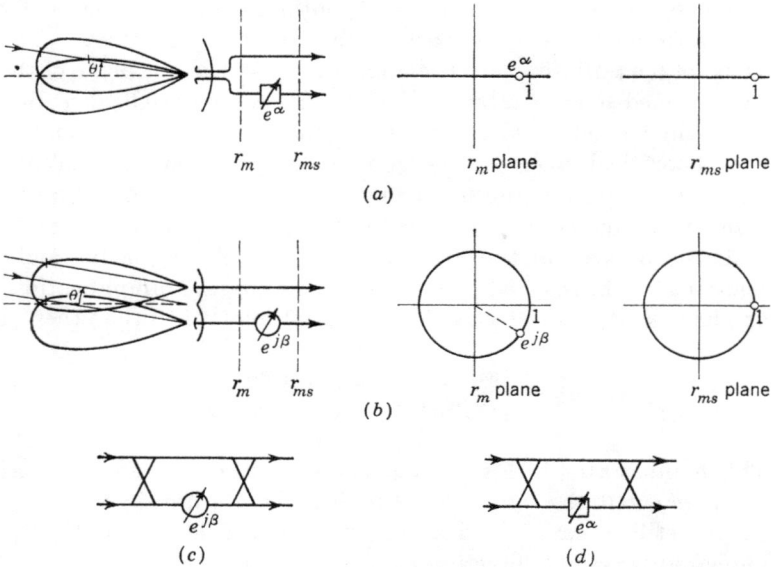

Fig. 3.15 Electronic boresight scanning: (a) amplitude sensing, (b) phase sensing, (c) phase scanner for amplitude- (or sum-and-difference) sensing functions, and (d) amplitude scanner for phase-sensing functions.

and b may be used directly with classes II and III since the boresight may be shifted from unity by inserting attenuation or phase shift, respectively. Class I, on the other hand, is totally unaffected by attenuation or phase shift since its boresight is indicated by a null. Nevertheless its boresight may be shifted, either by inserting an appropriate attenuation or phase shift ahead of the sum-and-difference network or by converting the sum-and-difference sensing function (which is an amplitude function) into an equivalent phase function and then converting back again as in Fig. 3.15c. The latter technique is illustrated in Fig. 3.16.

The parameters of importance in any boresighting application are (1) the indicated boresight direction, and (2) sensitivity of the indicated boresight. In all cases of electronic boresight scanning the indicated boresight is related to the scanning attenuation or phase shift through Eq. (3.8). The other parameter, boresight sensitivity, is the slope $\dfrac{\partial \, \mathrm{Re} \, \mathfrak{F}}{\partial u}\bigg|_{u=u_{ib}}$ of the angle-output function in the boresight direction. Its value depends upon the particular choice of the angle-detection function $\mathfrak{F}$. For the systems shown in Fig. 3.12 these reduce to

$$\frac{\partial \, \mathrm{Re} \, \mathfrak{F}}{\partial u}\bigg|_{u=u_{ib}} = \begin{cases} e^{\alpha}\rho'(u_{ib}) & \text{classes } I_A \text{ and II} \\ \phi'(u_{ib}) & \text{classes } I_P \text{ and III.} \end{cases} \tag{3.10}$$

A specific example of electronic boresight scanning as it might be used in a complete modern monopulse radar is illustrated in Fig. 3.16. It is basically a class IA monopulse system, probably the most common form in use today, modified for electronic scanning by the insertion of an electronic boresight scanner. The scanner has been placed beyond the TR switches so that it affects only the receiving characteristics of the radar. By using short slot couplers instead of magic T's, the entire r-f portion of the system, including the electronic scanner, can be built compactly into a pair of parallel wave guides with a common wall for coupling. The scanning phase shift β can be controlled mechanically by a movable dielectric in the wave guide, or all electronically by magnetizing a ferrite in the wave guide. The signals formed by the angle sensor are in-phase and have amplitudes that are the sum and the difference of the pattern amplitudes. After passing through the scanner they still remain in-phase, but their amplitudes involve both the sum and the difference:

$$\Sigma_s \triangleq \Sigma \cos \frac{\beta}{2} - \Delta \sin \frac{\beta}{2},$$

$$\Delta_s \triangleq \Delta \cos \frac{\beta}{2} + \Sigma \sin \frac{\beta}{2}. \tag{3.11}$$

Consider first the situation with no phase shift ($\beta = 0$). The scanner then has no effect on the system (other than to reverse the roles of the sum and the difference channels). The system operates as a true monopulse radar from which angle and range are obtained as functions of time; the signals from the scanner are just the sum and the difference shown before in Fig. 2.7 as func-

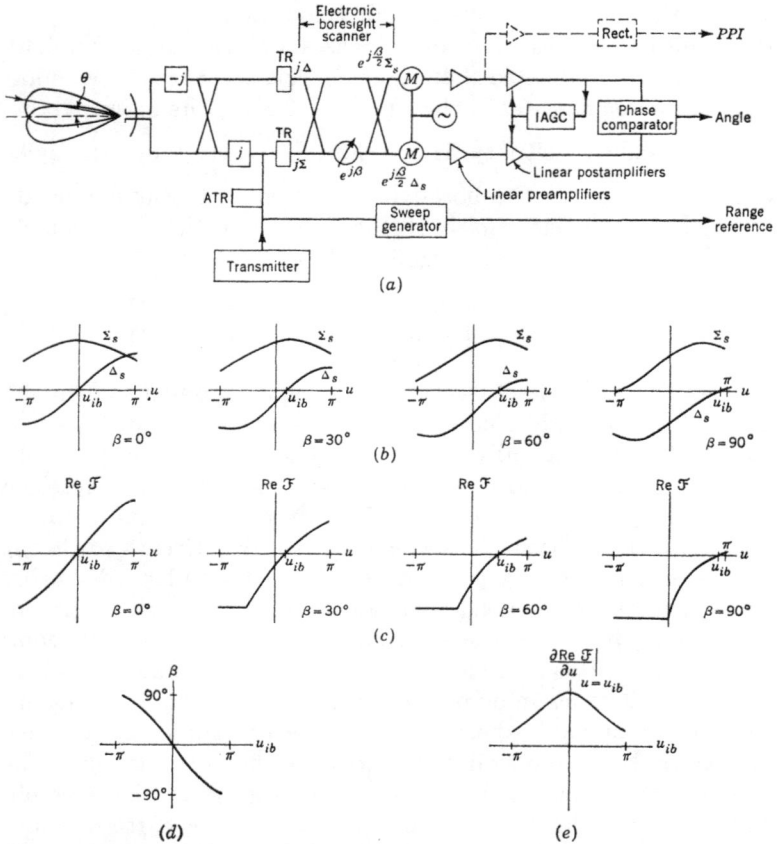

Fig. 3.16 A complete monopulse radar system with electronic boresight scanning: (a) schematic diagram of the radar, showing a PPI video output (dotted) from the sum channel in addition to the monopulse angle output, (b) the sum-and-difference pattern functions from the scanner, (c) angle output from the radar, (d) scanning phase shift β required to produce a given change in the indicated boresight direction, and (e) boresight sensitivity of the angle output function. All of the functions shown were calculated for a cosinusoidal aperture amplitude function.

tions of the generalized angle of arrival $\pi(d/\lambda) \sin \theta$, and again in Fig. 3.16b (within the half-power sum beamwidth). The sum pattern Σ_s for $\beta = 0$ represents also the pattern transmitted by the radar. The difference pattern Δ_s passes through zero in the direction of the actual boresight $u = 0$; hence the boresight direction indicated by the angle output from the radar is the actual boresight direction. The resulting angle-output function from the radar is shown in Fig. 3.16c. Now by varying the scanning phase β the effect of electronic scanning on the sum and the difference functions and on the angle-output function can be seen clearly. As β increases toward 90°, the indicated boresight moves out to the edge of the beam (the half-power point of the sum pattern). By varying β over the range $\pm 90°$ a target can be tracked over the entire width of the beam. The boresight direction indicated by electronic boresight scanning over the beamwidth is a single-valued, fixed function of the scanning phase shift, related as shown in Fig. 3.16d; hence the amount of scanning phase shift inserted offers an accurate measure of the direction of the indicated boresight. In comparing the performance of electronic boresight scanning with ordinary mechanical boresight scanning, it will be seen to have the disadvantage of variable sensitivity over the scanning range. This is evident in the changing slope of the angle-output function about the indicated boresight. Boresight sensitivity as a function of the indicated boresight direction is shown in Fig. 3.16e, in which it is seen to drop at the edges of the beam to about one-third its value at the center of the beam. If the signal-to-noise ratio is sufficiently high, this is not a particularly important disadvantage.

It may be of interest to note that a class I monopulse radar, such as that in Fig. 3.16, may also be used for PPI as well as for making precision angle-of-arrival measurements. If a part of the sum signal is bled off before being subjected to AGC and then amplified and detected (shown dotted in Fig. 3.16a), the video output will exhibit the variations in radar return that would be seen by a conventional search radar. In this way a monopulse radar may be used in a conventional search mode as well (electronic boresight scanner fixed at $\beta = 0$). It is even possible that the search and the electronic boresight-scanning modes could be operated simultaneously if the change in the sum pattern that occurs during electronic scanning could be made to have a negligible effect on the search video output.

One form of electronic boresight scanning that has actually been put into practice has been described by Sommer.[5] The system was part of a dual-plane missile-guidance seeker designed to track and home on active r-f sources. It is one of the two simplest forms of electronic boresight scanning, namely a class IIIP system with a phase scanner at one of the interferometer feeds (Fig. 3.17). The angle output of the system is converted into a mechanical rotation to drive a variable phase shifter in a closed servo loop. The phase shifter then tracks the target electronically, the angular position of the phase shifter giving a continuous indication of the electrical boresight, and hence of the target direction, relative to the mechanical boresight direction. It may be of interest to note that this is one of the few examples reported of a monopulse system developed to sense active sources only. By far the majority of applications have been in the field of radar.

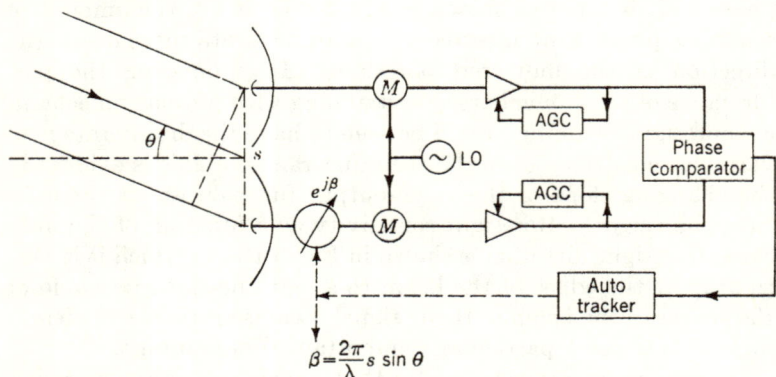

$$\beta = \frac{2\pi}{\lambda} s \sin \theta$$

Fig. 3.17 A phase-scanning missile-guidance seeker. (*After Sommer.*)

Upon examination of the scanning ratio in Eq. (3.7) it will be seen to lack symmetry about the indicated boresight (except when the scanning parameter α or β is zero) and hence will not satisfy the third postulate of monopulse. Strictly speaking, then, electronic boresight scanning is a form of pseudomonopulse since it does not satisfy the definition of monopulse. Here is an important example of a case in which the theory of monopulse has been used to develop a technique that does not satisfy the postulates on which that theory is based. This reemphasizes the assertion made earlier that the theory proposed can be of value in the analysis and synthesis of many systems that are related to, but may not be exactly, true monopulse as defined here.

4. Dual-plane monopulse systems

THE ESSENTIAL properties of some of the monopulse systems in use have been derived in Chap. 3 from the special theory. Although restricted, for simplicity, to two-dimensional monopulse operation in a single plane, the same properties also apply to three-dimensional monopulse operation in two orthogonal planes. Representative techniques for achieving dual-plane monopulse operation will now be described. As in the case of single-plane systems, the specific techniques to be described are merely illustrations of ways in which dual-plane systems may be realized physically, and hence are far from being exhaustive.

4.1 Three-channel techniques

The most direct extension from single- to dual-plane monopulse operation would be simply to combine two complete single-plane systems, one in azimuth and the other in elevation. This would require four antenna beams and four amplifying channels. Upon further consideration, however, it is evident that some simplification should be possible since the two systems could share one beam and one channel. A complete dual-plane system based upon this technique was developed early in the history of monopulse. Perhaps the first dual-plane monopulse radar system[1] (Sec. 1.4), it had three identical beams from which the signals received were amplified in separate channels for phase sensing and phase detection in azimuth and elevation, with a

69

fourth beam used only for transmission. A major disadvantage of such a system is that the antenna aperture is utilized inefficiently, part of the antenna being used only for azimuth reception, part for elevation reception, and part for transmission. The situation is even worse in the case of amplitude sensing since the receiving array of three squinted beams is inherently asymmetrical. An array of four beams, on the other hand, is completely symmetrical in any plane parallel to the two principal planes in both the cases of amplitude and phase sensing, as seen from Fig. 4.1. Furthermore, the full antenna aperture can be utilized for both azimuth and elevation sensing by comparing signals from pairs of beams, the right and left pairs compared in azimuth and the upper and lower pairs compared in elevation.

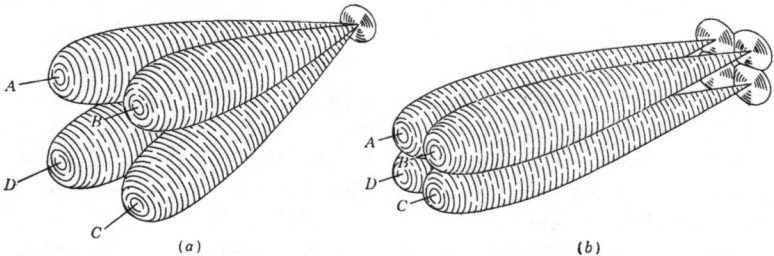

Fig. 4.1 Dual-plane sensing with four beams: (a) amplitude sensing, and (b) phase sensing.

This can be of the utmost importance whenever antenna size is limited. With four beams the full aperture can be utilized on transmission as well as reception if duplexing provisions are included to feed all four equally during transmission. This points once again toward the use of sum-and-difference sensing. On transmission the transmitter can feed the sum of all four beams, while on reception the received signal from the sum of all four beams can be used as the common reference channel shared by both planes. This, plus the fact that dual-plane monopulse has been used primarily in boresighting applications, has led almost exclusively to the use of class I for dual-plane monopulse.[2,4,7] Therefore only class I systems will be treated here. Some of the characteristics of class I operation are described in more detail in Chap. 6.

In either of the cases of pure amplitude or phase sensing shown in Fig. 4.1 the signals from all four beams added in-phase will produce a symmetrical sum pattern. Subtracting the signals

from the two lower beams C and D taken together from those of
the two upper beams A and B taken together will result in an
elevation difference signal, while subtracting signals of the two
left-hand beams B and C (as seen from the antenna) from those
of the two right-hand beams A and D will result in an azimuthal
difference signal. In the case of pure amplitude sensing (Fig.
4.1a), for example, the sum and the difference patterns are

$$\Sigma(u) = \tfrac{1}{2}[P_A(u) + P_B(u) + P_C(u) + P_D(u)],$$
$$\Delta_{el}(u) = \tfrac{1}{2}\{[P_A(u) + P_B(u)] - [P_C(u) + P_D(u)]\}, \quad (4.1)$$
$$\Delta_{az}(u) = \tfrac{1}{2}\{[P_A(u) + P_D(u)] - [P_B(u) + P_C(u)]\}.$$

The factor $\tfrac{1}{2}$ is necessary for conservation of power in a lossless
network. On the boresight axis $u = 0$, $P_A = P_B = P_C = P_D$, so

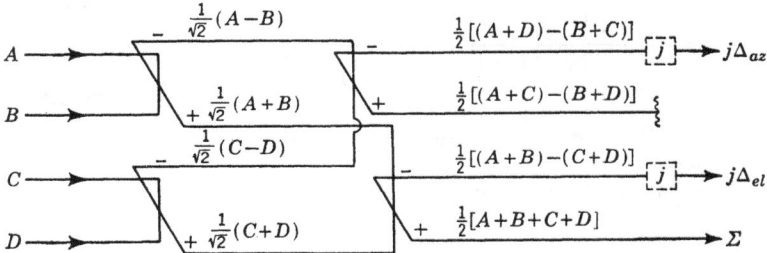

Fig. 4.2 A dual-plane sum-and-difference network.

that $\Delta_{el}(0) = \Delta_{az}(0) = 0$ and $\Sigma(0) = 2P_A(0)$. The total power
into the network, assumed to be matched, is $4P_A{}^2(0)$ on the boresight, which must equal the power $\Sigma^2(0)$ out of it. A set of
equations equivalent to those of Eq. (4.1) can be written also
for pure phase sensing.

A network for realizing the dual-plane sum-and-difference
signals of Eq. (4.1) is shown in Fig. 4.2. The signal voltages
have been designated simply A, B, C, and D, for either amplitude
or phase sensing. For amplitude sensing a 90° phase shift (shown
dotted) is inserted in the difference channels in order that the
ratio will be of the form $j\Delta(u)/\Sigma(u)$ for either amplitude or phase
sensing. This network, when excited by either set of four beams
in Fig. 4.1, is the dual-plane extension of the single-plane additive
angle sensor in Fig. 3.2. A third difference signal the difference
between the two pairs of diagonal beams, is also produced, but
since its information is redundant its power may be dissipated in
a matched load.

A dual-plane extension of the single-plane class I angle detector is illustrated in Fig. 4.3. The normalizing IAGC voltage is generated by the sum-amplifier signal and applied to the sum amplifier and both difference amplifiers. The heterodyning signal from the LO has to be applied to three mixers (plus an AFC mixer in most practical receivers); hence it is divided twice as shown.

When used as a radar, range information is available from any of the three channels. In the case of the two difference-channel signals, however, range information vanishes on the boresight with the vanishing of the difference amplitudes. The sum

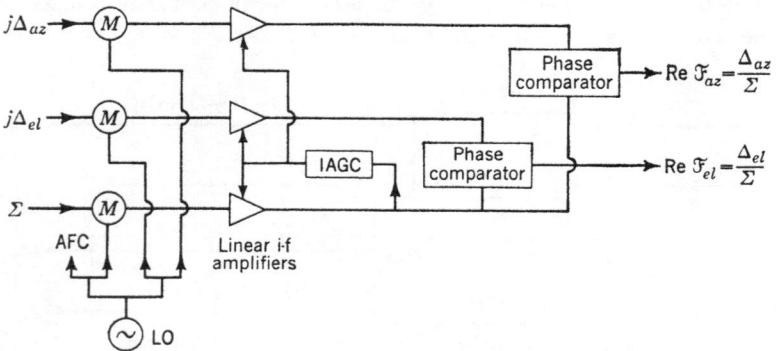

Fig. 4.3   A dual-plane three-channel class I angle detector.

signal, on the other hand, is a maximum in this critical direction. For this reason range information is always taken from the sum channel. Since the difference signals indicate a displacement of the radiating source from the boresight axis, they represent an angular error in boresighting applications. Consequently the sum and the difference signals of a class I monopulse system are frequently referred to as *range* and *error* signals, respectively.

4.2   Two-channel techniques

It is possible to reduce the number of separate channels required from three to two by combining the two difference signals in phase quadrature into a single complex signal. The complex difference signal can then be amplified in a single channel, after which the two original difference signals can be separated again by resolving the complex signal into its components in-phase and 90° out-of-phase with respect to the sum signal. This technique reduces the number of difference amplifiers required but does so at the expense of tighter phase tolerances. It is still true that the boresight direction will be indicated accurately by a null in the amplitude of the difference signal independently of the amplitude or phase characteristics of the amplifiers, which is the principal advantage of class I. Off of the boresight any inadvertent shift of phase in the complex difference amplifier relative to that of the sum amplifier will result in an indicated azimuth and elevation that is in truth a combination of the two. In boresight tracking applications any unwanted phase shift would result in cross correction of the azimuth error by the elevation error signal, and vice versa. A phase error as great as 90° would cause a complete interchange of azimuth and elevation error signals, thereby rendering the system useless for tracking; an elevation error signal would cause an azimuth correction, and vice versa. With two separate error channels, however, as in the case of the three-channel technique described in the preceding section, a phase error causes only a decrease in sensitivity (but with a catastrophic reversal of sense if the phase error becomes greater than 90°), as will be shown in Sec. 6.2, but it does not cross-couple the error signals.

Four-beam sensing

The complex difference signal can be obtained directly from the three-channel sensor described earlier in Figs. 4.1 and 4.2. All

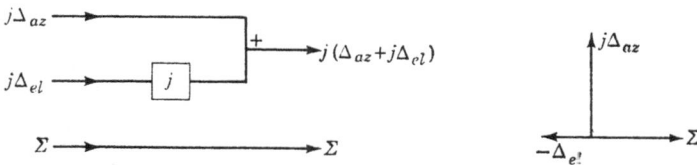

Fig. 4.4 Circuit for reducing three-channel information to two.

that needs to be done is to add the two difference signals after shifting the phase of one of them by 90°. This can be effected as shown in Fig. 4.4.

With the angle information contained in relative phase as well as in relative amplitude of the difference signals it becomes particularly important to reduce to a minimum the effect of the nonsensing function. In the case of amplitude sensing this means that

all four beams should have coincident phase centers, and in the
case of phase sensing that all four beams should be patterns of
revolution, i.e., should have cylindrical symmetry.   Under these
conditions the proper phase relation between the sum and the
complex difference signal is assured.

Two-beam sensing

Dual-plane angle sensing can be realized directly from a single
pair of beams.   The simplest possible form of dual-plane angle
sensor is obtained[4] simply by squinting the two beams of a single-
plane interferometer as indicated in Fig. 4.5.   The angle-sensing

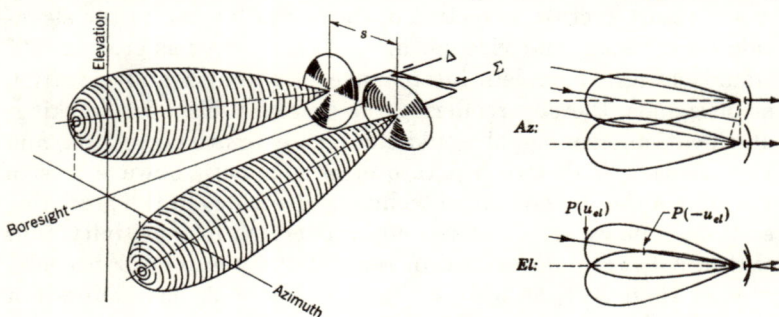

Fig. 4.5 A two-beam sensor.   The azimuth angle-sensing function is
phase, while the elevation angle-sensing function is amplitude.

function in the principal azimuth plane is phase, and in the principal elevation plane is amplitude.   This still constitutes simple
sensing in the two principal planes, consistent with the special
theory, even though it is complex sensing in any other plane
because of its unsymmetrical character.   A source displaced off
of the boresight axis will induce voltages that differ in phase
according to its azimuth displacement and that differ in amplitude according to its elevation displacement (Fig. 4.6a).   On
the boresight axis the two voltages are equal and in phase.   The
difference signal then passes through zero on the boresight.   With
the source in one of the two principal planes the angle sensor
operates simply as an ordinary single-plane sensor.   In the
principal azimuth plane the signals received are of equal amplitude (Fig. 4.6b) and the sensor reduces to the single-plane equivalent of an interferometer.   In the principal elevation plane the
signals received are in-phase (Fig. 4.6c) and the sensor reduces
to the single-plane equivalent of a pair of squinted beams.   For

small displacements from either principal plane the components of the complex difference signal parallel and perpendicular to the sum signal approximate, respectively, the elevation and azimuth error signals:

$$\Delta_\parallel(u_{el}) \approx P(u_{el}) - P(-u_{el}),$$

and

$$\Delta_\perp(u_{az}) \approx P(0)\phi(u_{az}). \tag{4.2}$$

Near the boresight, then, the sum and the two orthogonal components of the complex difference signal from the two-beam sensor

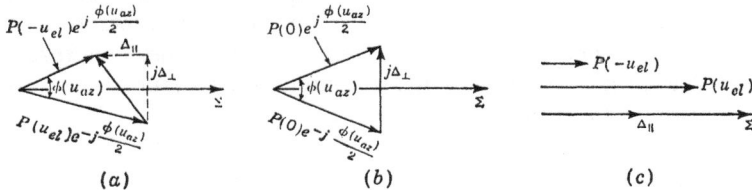

(a)   (b)   (c)

Fig. 4.6 Signals produced by the two-beam sensor in Fig. 4.5 when (a, the source is displaced from both principal planes, (b) the source lies in the azimuth principal plane, and (c) the source lies in the elevation principal plane.

are first-order approximations to those from the four-beam sensor. The approximation becomes poorer for sources farther from the boresight, however, because both azimuth and elevation become intermixed in each of the two components.

Angle detection

In either of the two principal planes the angle information from the four- and the two-beam sensors is in exactly the same form;

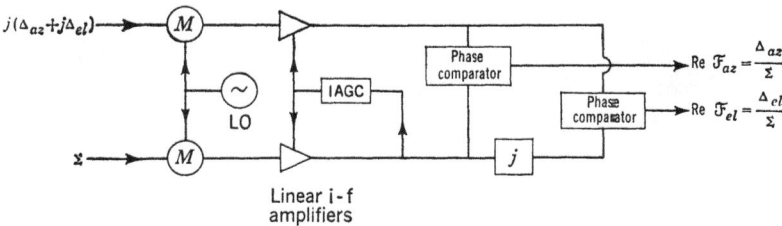

Fig. 4.7 A dual-plane two-channel class I angle detector.

hence it can be detected by the same angle detector. In both cases the complex difference signal can be resolved into its two components by an angle detector such as that in Fig. 4.7. The first phase comparator, being sensitive only to the perpendicular component, detects the azimuth angle-output function. **The** phase comparator following the 90° shift of phase is sensitive only to the in-phase component; hence it detects the elevation angle-output function.

In any plane other than one of the two principal planes the angle information from the four- and the two-beam sensors can differ considerably. In fact, since the two-beam angle information in any other plane is unsymmetrical about the boresight, it does not represent true monopulse information. The two-beam system can still be used in the usual way for boresight tracking, however, because it will always develop a suitable error signal with correct sense. As the boresight is brought onto the target, it will approach a first-order approximation to true monopulse operation in both planes.

5. Monopulse antenna principles

THE ANTENNA is the heart of monopulse. It is the link between the radiating source and the angle detector, sensing the angle of arrival by virtue of its configuration. Once the basic form of the angle detector is chosen from among the relatively few forms in use, the principal control of the monopulse angle characteristics at the present time rests in the design of the antenna. By choosing the proper distribution of radiating fields in the antenna aperture it is possible to achieve control of the angle-output function. For example, it can be linearized over the angular-scanning range, or its boresight sensitivity can be maximized, by the aperture synthesis techniques described in the next chapter. In any case an understanding of the principles of monopulse antennas is essential to a full understanding of the operation of monopulse systems.

5.1 Relationship between aperture and pattern functions

The monopulse concept was defined in terms of the ratio of antenna pattern functions, and the various monopulse techniques were described in terms of these functions. But to control the pattern functions to achieve a prescribed angle output it is necessary to synthesize the required current distribution in the aperture. Synthesis, and the practical problem of physical realization of the synthesized distribution required to achieve a prescribed angle output, are the principal problems of monopulse-antenna design. Therefore a brief review will be given of the elementary Fourier relationship between aperture and pattern

77

functions from which the general properties of monopulse aperture functions will be derived. Admittedly this representation of the relationship is only an approximation.* Its virtue lies in the fact that it is usually a good approximation to the true relationship, leading to simple conditions on the aperture functions.

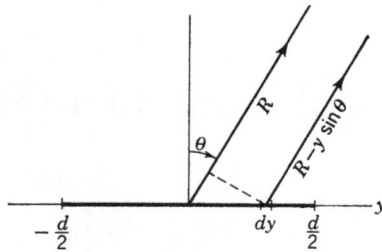

Fig. 5.1 A radiating antenna aperture.

Consider an antenna with a one-dimensional, monochromatic current distribution $\alpha(2y/d)e^{j\omega t}$ across the aperture. The distant field radiated by an element of length dy located at y (Fig. 5.1) will be proportional to

$$d\mathcal{P} = \frac{2}{d}\,\alpha\left(\frac{2}{d}y\right)\frac{e^{j\omega\left(t - \frac{R - y\sin\theta}{c}\right)}}{R}\,dy.$$

The total field radiated in any given direction is then a linear superposition of the fields radiated in that direction by the entire aperture,

$$\mathcal{P}\left(\pi\frac{d}{\lambda}\sin\theta\right) = \frac{2}{d}\int_{-d/2}^{d/2}\alpha\left(\frac{2}{d}y\right)e^{j\frac{2\pi}{\lambda}y\sin\theta}\,dy.$$

The radial and time dependence appear as constant factors and hence were omitted for simplicity. The dependent variables and proportionality constant were chosen in such a way as to reduce this relationship to a simple form after the usual change of variables,

$$u \triangleq \pi\frac{d}{\lambda}\sin\theta,$$

and
$$v \triangleq \frac{2}{d}y. \tag{5.1}$$

---

* H. G. Booker and P. C. Clemmow, The Concept of Angular Spectrum of Plane Waves, and Its Relation to that of Polar Diagram and Aperture Distribution, *JIEE (London)*, pt. III, vol. 97, pp. 11–17, January, 1950.

Variables u and v are the generalized pattern and aperture coordinates, respectively. Then

$$\mathcal{P}(u) = \int_{-1}^{1} \mathcal{A}(v)e^{iuv}\, dv. \tag{5.2}$$

The pattern function is recognized as being the Fourier transform of the aperture function [$\mathcal{A}(v) = 0$ for $|v| > 1$; hence $\mathcal{A}(v)$ is defined on the entire interval $-\infty < v < \infty$]. Consequently the aperture function is the inverse Fourier transform of the pattern function:

$$\mathcal{A}(v) = \frac{1}{2\pi} \int_{-\infty}^{\infty} \mathcal{P}(u)e^{-iuv}\, du. \tag{5.3}$$

Given any arbitrary aperture function its corresponding pattern function may be obtained exactly from Eq. (5.2). For example, in the simplest case of a uniform aperture function the complete pattern will extend over a range of u from $-\infty$ to ∞, as illustrated in Fig. 5.2. Only that portion of the pattern function

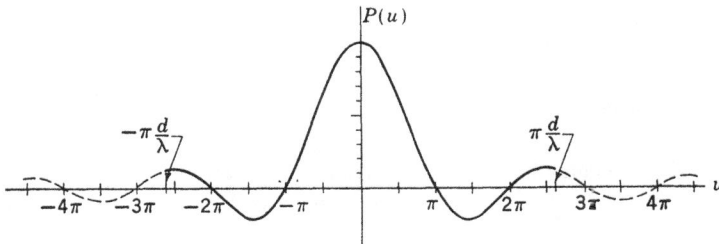

Fig. 5.2 The radiation pattern of a uniform line source.

lying in the interval $-\pi(d/\lambda) \leq u \leq \pi(d/\lambda)$, i.e., $-1 \leq \sin\theta \leq 1$, is observable over the physical range of angles, however, even though the pattern function is defined over the entire real axis. For the portion of the pattern function outside of this interval the angle of arrival is complex; this portion has been referred to as the *invisible* region. The inverse problem of synthesizing the exact aperture function required to produce a given pattern function has no solution, in general. The difficulty lies in the fact that the aperture function must vanish everywhere outside of the aperture, whereas the inverse transform [Eq. (5.3)] does not, except for certain very special cases. Given an arbitrary pattern function over the visible range of angles, it is possible to find an aperture function that will approximate it as closely as desired, however, by choosing the invisible portion of the pattern function properly. This is the theoretical basis for the possibility of supergain antennas. An ingenious and novel approach to the problem of aperture synthesis has been given by Woodward and Lawson* in which they obtain the aperture function by linear superposition of a continuous spectrum of plane waves. Over the visible region the spectrum is distributed according to the required pattern function. The invisible portion, interpreted as arising from a spectrum of plane waves at *complex angles* of incidence, was chosen in such a way as to reduce the fields outside of the aperture itself to zero. The aperture function was then expressed as a sum of plane waves over the entire complex

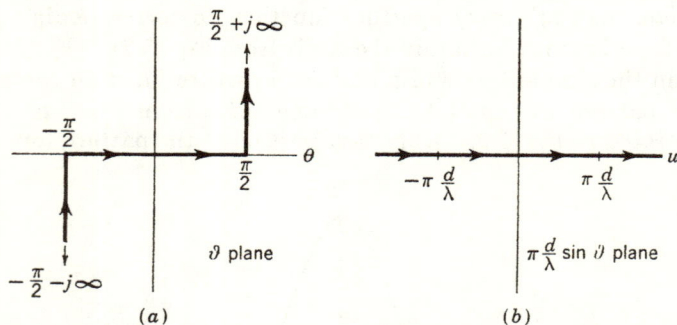

Fig. 5.3 The only angle-of-arrival contour with physical significance: (*a*) the ϑ-plane contour, and (*b*) the corresponding $\pi(d/\lambda) \sin \vartheta$-plane contour.

spectrum of angles. It was shown that the integral must be restricted to the ϑ-plane contour shown in Fig. 5.3*a* in order that it will always remain bounded. This corresponds to the entire real axis in the complex $\pi(d/\lambda) \sin \vartheta$ plane (Fig. 5.3*b*). By showing that the power in the visible and invisible portions of the pattern function corresponds, respectively, to the real and reactive power passing through the aperture, they arrived at the important conclusion that the visible portion of the pattern function represents the total power radiated by the aperture, whereas the invisible portion represents energy stored about the aperture. When the reactive power becomes large relative to the radiated power, the antenna becomes a high-Q circuit element with an

* P. M. Woodward and J. D. Lawson, The Theoretical Precision with Which an Arbitrary Radiation Pattern May Be Obtained from a Source of Finite Size, *JIEE* (*London*), pt. III, vol. 95, pp. 363–370, September, 1948.

attendant narrow bandwidth, high ohmic losses, and low construction tolerances. These are the physical limitations inherent to the synthesis of arbitrary pattern functions.

5.2 Monopulse antenna-function criteria

The monopulse postulates impose some fundamental limitations on the form that the pattern and aperture functions can take for simple sensing. These will be derived here.

Two forms of comparison [Eq. (2.4)] of the received signals were admitted by the monopulse postulates. Both forms were expressed in terms of the multiplicative sensing ratio

$$r_m(u) = \frac{\mathcal{P}(u)}{\mathcal{P}(-u)}.$$

It is this ratio that must be examined.

Pattern functions

The multiplicative sensing ratio, expressed in terms of the pattern amplitude and phase functions, is

$$r_m(u) = \frac{P(u)e^{j\varphi(u)}}{P(-u)e^{j\varphi(-u)}}.$$

The sensing pattern function may be either amplitude $P(u)$ or phase $\varphi(u)$. In either case the two nonsensing pattern functions must be identical so that they cancel in the ratio. A necessary and sufficient condition for pure amplitude or phase sensing, then, is that the nonsensing pattern function be an even function about the boresight axis, i.e., for pure amplitude sensing

$$\varphi(u) \underset{u}{\equiv} \varphi(-u), \tag{5.4a}$$

and for pure phase sensing,

$$P(u) \underset{u}{\equiv} P(-u). \tag{5.4b}$$

Aperture functions

Conditions [Eq. (5.4)] on the nonsensing pattern function place certain restrictions on the form that the aperture functions can take. One general condition may be noted directly, namely, that the aperture functions corresponding to the mirror-image pattern functions must themselves be mirror images. This is seen by a simple change of variables in the Fourier integral:

$$\mathcal{P}(-u) = \int_{-1}^{1} \mathcal{Q}(v)e^{-iuv}\,dv$$

$$= \int_{-1}^{1} \mathcal{Q}(-v)e^{iuv}\,dv.$$

Let the complex aperture function $\mathcal{Q}(v)$ be represented by

$$\mathcal{Q}(v) \triangleq A(v)e^{i\alpha(v)}. \tag{5.5}$$

Then the conditions for pure amplitude and phase sensing given by Eqs. (5.4a and b) are, respectively,

$$\frac{\int_{-1}^{1} A(v)\sin\,[\alpha(v) + uv]\,dv}{\int_{-1}^{1} A(v)\cos\,[\alpha(v) + uv]\,dv} \underset{u}{\equiv} \frac{\int_{-1}^{1} A(v)\sin\,[\alpha(v) - uv]\,dv}{\int_{-1}^{1} A(v)\cos\,[\alpha(v) - uv]\,dv}, \tag{5.6a}$$

and

$$\left| \int_{-1}^{1} A(v)e^{i[\alpha(v)+uv]}\,dv \right| \underset{u}{\equiv} \left| \int_{-1}^{1} A(v)e^{i[\alpha(v)-uv]}\,dv \right|. \tag{5.6b}$$

Equation (5.6a) equates the tangents of $\varphi(u)$ and $\varphi(-u)$. The problem now is to establish conditions on the amplitude $A(v)$ and phase $\alpha(v)$ that satisfy these identities.

Rather than to try to obtain general conditions on both $A(v)$ and $\alpha(v)$, the problem can be simplified immensely by recognizing that in actual practice the aperture phase function $\alpha(v)$ is limited to a fairly specific form. In Sec. 3.2 it was pointed out that sensing is usually performed by squinting two ordinary beams at a small angle away from the boresight axis, or by displacing the phase centers of the two beams. Squinted beams are generated by a pair of waves traveling with constant phase velocity in opposite directions in the aperture; hence the aperture phase function is usually linear:

$$\alpha(v) = -u_s v. \tag{5.7}$$

The slope of the phase function, denoted by $-u_s$, is related to the squint angle θ_s by

$$u_s \triangleq \pi \frac{d}{\lambda} \sin\,\theta_s \tag{5.8}$$

through the phase velocity $c/\sin\,\theta_s$. Displaced phase centers, on the other hand, are produced by a linear aperture phase function that is constant:

$$\alpha(v) \equiv \text{const.} \tag{5.9}$$

Linear phase functions are generally recognized as being of fundamental importance to radar-antenna design because they result in nearly maximum realizable gain from a given size of aperture.* Henceforth it will be assumed that the aperture phase functions are linear for amplitude sensing, and constant for phase sensing. General conditions on the amplitude functions will now be derived on the basis of this assumption.

Consider first the case of pure amplitude sensing. With a linear phase function the condition [Eq. (5.6a)] on the pattern function is

$$\frac{\int_0^1 A_o(v) \sin (u_s - u)v \, dv}{\int_0^1 A_e(v) \cos (u_s - u)v \, dv} \underset{u}{\equiv} \frac{\int_0^1 A_o(v) \sin (u_s + u)v \, dv}{\int_0^1 A_e(v) \cos (u_s + u)v \, dv}, \tag{5.10}$$

where $A_o(v)$ and $A_e(v)$ are the odd and even parts of $A(v)$. It is evident by inspection that a sufficient condition satisfying this identity is either

$$A_o(v) \equiv 0, \tag{5.11a}$$

or

$$A_e(v) \equiv 0, \tag{5.11b}$$

for any u_s. It will now be argued that Eq. (5.11) is a necessary as well as a sufficient condition. Equation (5.10) requires that the ratio itself be an even function of u. This is satisfied if, and only if, (1) either the odd or the even parts of the two integrals vanish identically, which leads to the trivial case of $A(v) \equiv 0$, or (2) the ratio is a constant, which leads to the significant case of

$$\int_0^1 A_o(v) \sin (u_s - u)v \, dv - c \int_0^1 A_e(v) \cos (u_s - u)v \, dv \underset{u}{\equiv} 0, \tag{5.12}$$

where c is the constant. If Eq. (5.12) is integrated over the interval $-u' \leq u \leq u'$, where u' is arbitrary, we obtain a new identity in u':

$$\int_0^{u'} \left[\int_0^1 A_o(v) \sin u_s v \cos uv \, dv - c \int_0^1 A_e(v) \cos u_s v \cos uv \, dv \right] du \underset{u'}{\equiv} 0.$$

* S. Silver, "Microwave Antenna Theory and Design," McGraw-Hill Book Company, Inc., New York, 1949, p. 414.

Interchanging the order of integration reduces this to

$$\int_0^1 [A_o(v) \sin u_s v - cA_e(v) \cos u_s v] \frac{\sin u'v}{v} \, dv \underset{u'}{\equiv} 0,$$

which can be satisfied for all u' only if

$$A_o(v) \sin u_s v - cA_e(v) \cos u_s v \underset{u}{\equiv} 0.$$

If the identity [Eq. (5.12)] is first differentiated with respect to u and the preceding process repeated, we obtain

$$A_o(v) \cos u_o v - cA_e(v) \sin u_s v \underset{u}{\equiv} 0.$$

But the last two identities are mutually consistent only if c is either zero or infinite. If c is zero, then $A_o(v)$ must vanish identically, or if c is infinite, then $A_e(v)$ must vanish identically. Therefore, it is concluded that *a necessary and sufficient condition for amplitude sensing with linear aperture phase is that the aperture amplitude be either an even or an odd function about the center of the aperture.*

Consider next the case of pure phase sensing. With a constant aperture phase the condition [Eq. (5.6b)] for phase sensing becomes

$$\left| \int_{-1}^1 A(v)e^{juv} \, dv \right| \underset{u}{\equiv} \left| \int_{-1}^1 A(v)e^{-juv} \, dv \right|.$$

It is evident by inspection that this is satisfied by any amplitude $A(v)$ since the two integrals are always complex conjugates of each other. Therefore, *for phase sensing with a constant aperture phase the aperture amplitude may be any arbitrary function.* A consequence of constant phase in the aperture is the fact that the pattern phase must be an odd function. Therefore the pattern function will always have an even amplitude and an odd phase about the boresight axis.

5.3   Phase centers

The concept of phase center of an antenna arises frequently in monopulse theory. The aperture-function criterion for pure amplitude sensing implies coincidence of phase centers, and the most practical case of pure phase sensing consists of interference between displaced phase centers. Therefore the conditions for the existence of phase centers and determination of their location will be derived here.

The phase pattern of many conventional antennas is essentially circular within the main beam. The center of this circle is defined as the phase center of the pattern. It is an actual geometrical point in space, fixed relative to the antenna structure. Since the two patterns of a single-plane monopulse antenna are mirror images about the boresight direction, the phase centers must be located symmetrically with respect to that reference axis. If the aperture is assumed to lie on the line through the phase centers at $\pm y_{pc}$ (Fig. 5.4a), then the location of the phase centers may be used to define the physical location of the aperture in space. This points up a fact that is frequently misunderstood, namely, that there is no unique aperture. Any surface about an

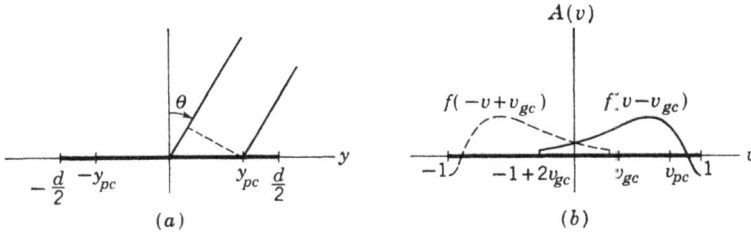

Fig. 5.4 (a) A monopulse antenna aperture with phase centers at $\pm y_{pc}$, and (b) an antenna aperture function (solid) and its image (dotted).

antenna may be taken as the radiating aperture, provided that the tangential components of the actual fields over that surface are taken as the equivalent current distribution in that aperture.[*] In any practical system the equivalent current distribution does not vanish completely outside of the area taken as the aperture, although it is usually small. It is usually assumed, however, that it does vanish outside of the aperture, and since the nonvanishing portion of the amplitude function differs but little over planes parallel but close to the antenna, it is permissible to choose the aperture to lie in a plane through the phase centers. In general the phase centers in the azimuth plane appear at a different location than those in the elevation plane,[†] which means that the aperture so defined will lie in a different plane for

[*] S. A. Schelkunoff, "Electromagnetic Waves," D. Van Nostrand Co., New York, 1943, p. 158.

[†] D. Carter, Phase Centers of Microwave Antennas, *IRE Trans. PGAP,* vol. AP-4, pp. 597–600, October, 1956.

the azimuth patterns than for the elevation patterns. The advantage of defining the location of the aperture to lie in the plane of the phase centers lies in the simplicity of the resulting phase functions. If a phase center lies at $\pm y_{pc}$, then its phase function relative to the origin is simply

$$\varphi(u) = \pm \frac{2\pi}{\lambda} y_{pc} \sin \theta$$

$$= \pm u v_{pc}.$$

(5.13)

Thus a phase center can exist in the aperture if, and only if, the pattern phase is linear.

The existence and location of phase centers for amplitude sensing are obtained directly from the fact that the aperture function must be either an even or an odd function about the center of the aperture [Eq. (5.11)]. The complex pattern function is then

$$\mathcal{P}(u) = \begin{cases} 2 \int_0^1 A_e(v) \cos (u - u_s)v \, dv \\ j2 \int_0^1 A_o(v) \sin (u - u_s)v \, dv, \end{cases}$$

which is always either purely real or purely imaginary. Since the pattern phase is always constant, it is concluded from Eq. (5.13) that it always does have a phase center, and that the phase center is located at the origin, i.e., $v_{pc} = 0$. Its image, then, also has a phase center located at the origin, which confirms the assertion made earlier that phase centers must be coincident for amplitude sensing.

The question of existence and location of phase centers in the case of phase sensing cannot be answered as directly because of the fact that both the odd and even parts of the aperture function are arbitrary. It will now be proved that for a constant phase in the aperture a necessary and sufficient condition for the existence of a phase center at an arbitrary point in the aperture is that $A(v)$ be an even function about that point.

The aperture amplitude may vanish over a portion of the aperture itself, in general, in addition to vanishing everywhere outside of the aperture. Let the nonvanishing portion of the aperture amplitude be represented by an arbitrary function $f(v - v_{gc})$, where v_{gc} is its geometrical center (Fig. 5.4b). The tangent of the pattern phase function is then

$$\tan \varphi(u) = \frac{\int_{-1}^{1} A(v) \sin uv \, dv}{\int_{-1}^{1} A(v) \cos uv \, dv}$$

$$= \frac{\cos uv_{gc} \int_{0}^{1-v_{gc}} f_o(x) \sin ux \, dx + \sin uv_{gc} \int_{0}^{1-v_{gc}} f_e(x) \cos ux \, dx}{\cos uv_{gc} \int_{0}^{1-v_{gc}} f_e(x) \cos ux \, dx - \sin uv_{gc} \int_{0}^{1-v_{gc}} f_o(x) \sin ux \, dx}.$$

Again, $f_o(x)$ and $f_e(x)$ are the odd and even parts of $f(x)$. After dividing the numerator and denominator by $\cos u_{gc} \int_{0}^{1-v_{gc}} f_e(x) \cos ux \, dx$, the right-hand side will be recognized as the tangent of a sum of two angles; hence

$$\varphi(u) = uv_{gc} + \tan^{-1} \left[\frac{\int_{0}^{1-v_{gc}} f_o(x) \sin ux \, dx}{\int_{0}^{1-v_{gc}} f_e(x) \cos ux \, dx} \right]. \qquad (5.14)$$

It remains to determine the conditions under which this is a linear function of u. By inspection it is evident that a sufficient condition is that $f_o(x)$ vanish, the arc tangent then being zero. This will now be shown to be a necessary condition as well, in view of the fact that the other obviously sufficient condition, $f_e(x) \equiv 0$, leads to the trivial solution of a pattern amplitude function that vanishes on the boresight. For a linear $\varphi(u)$ the arc tangent itself must be linear; i.e., of the form

$$\tan^{-1} \left[\frac{\int_{0}^{1-v_{gc}} f_o(x) \sin ux \, dx}{\int_{0}^{1-v_{gc}} f_e(x) \cos ux \, dx} \right] \equiv_u ua + b,$$

where a and b are constants to be determined. Hypothesize that neither $f_o(x)$ nor $f_e(x)$ vanish identically. Then b must be zero since the arc tangent is an odd function of u. If $a \neq 0$, then the identity must hold for $u = n\pi/a$, $n = 1, 2, 3, \ldots$, which would require that the numerator vanish for all integral values of n. But this would require that $f_o(x)$ vanishes identically since it cannot be orthogonal to all of the functions $\sin n\pi x/a$, in contradiction to the hypothesis. Similarly for $f_e(x)$ if we choose $u = (n + \frac{1}{2})\pi/a$, $n = 1, 2, 3, \ldots$. If, on the other hand, $a = 0$, then the identity requires that the numerator, and hence $f_o(x)$, must vanish, again in contradiction to the hypothesis. Therefore it is concluded that the hypothesis is false; either $f_o(x)$

or $f_e(x)$ must vanish identically. Since $f_e(x) \equiv 0$ is trivial, the only significant condition for existence of a phase center is that the aperture amplitude be an even function about its geometrical center. Then the arc tangent is zero, from which it is concluded that the phase center is located at the geometrical center.

5.4   Angled-boresight monopulse

The entire theory of monopulse has been developed under the implicit assumption that the boresight lies in the direction broadside to the plane of the aperture. It will be shown in this section that the theory is much more general and that, in fact, it applies to any arbitrary direction of the boresight. The key to the generalization lies in the choice of the generalized pattern

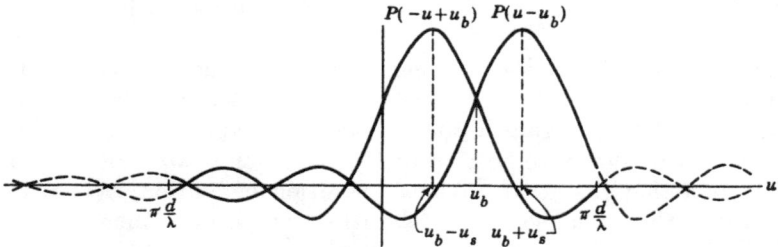

Fig. 5.5   Antenna patterns for angled-boresight monopulse operation.

coordinate u rather than in the angle of arrival itself as the independent variable. When expressed as a function of u, the pattern may be translated without distortion to any position on the u axis simply by an additional linear phase shift of the aperture function. This is illustrated in Fig. 5.5 for a uniform aperture amplitude. When the original pattern $\mathcal{P}(u)$ is translated along the u axis by the generalized boresight angle u_b, related to the physical boresight angle θ_b by

$$u_b \triangleq \pi \frac{d}{\lambda} \sin \theta_b, \tag{5.15}$$

it becomes

$$\mathcal{P}(u - u_b) = \int_{-1}^{1} A(v) e^{j[\alpha(v) + (u - u_b)v]} \, dv,$$

obtained physically by shifting the phase of the original aperture function linearly by $u_b v$. If the original pattern is squinted away from the boresight direction by u_s (Fig. 5.5), then the translated pattern will be squinted away from the angled-boresight direction u_b by an amount u_s. By translating both pattern functions from the broadside direction $u = 0$ to the angled-boresight direction $u = u_b$, all of the monopulse relationships will be preserved without distortion as a function of u about the *angled-boresight axis*, provided that the multiplicative sensing ratio is redefined to be

$$r_m(u) \triangleq \frac{\mathcal{P}(u - u_b)}{\mathcal{P}(-u + u_b)}. \tag{5.16}$$

This reduces to the usual definition [Eq. $(2.4a)$] when $u_b = 0$.

It must be clearly understood that the pattern functions, and hence the resulting monopulse relationships, remain undistorted only because they are expressed as functions of the generalized pattern coordinate. When expressed directly as functions of angle of arrival θ this is no longer true; e.g., the patterns will not be mirror images about the angled-boresight axis. Symmetry about the boresight, the property that led to the adoption of the last two postulates of monopulse, is maintained only when the patterns are expressed as functions of u.

5.5   Coupling between feeds

It has been tacitly assumed throughout that the signals induced in a monopulse antenna are independent of each other. In actual fact, however, the received signals are interdependent because of mutual coupling between feeds. A portion of the incident power received by any one feed is reradiated (e.g., half of the power received by a matched antenna is reradiated), part of which is coupled into each of the other feeds. The question then arises as to the effect of mutual coupling on the operation of a monopulse system. It will be shown here that the sole effect is to reduce the sum-and-difference sensing function by a constant factor, provided that the coupling is symmetrical.

In the simplest case, that of single-plane sensing, the coupled voltages induced in the two feeds are a linear combination of the two uncoupled voltages $P_A(u)$ and $P_B(u)$ (the argument will be restricted to pure amplitude sensing for the moment, but the results can be seen to apply equally well to pure phase sensing):

$$\begin{aligned}
P_{A_c}(u) &= P_A(u) + kP_B(u), \\
P_{B_c}(u) &= kP_A(u) + P_B(u).
\end{aligned} \tag{5.17}$$

and

The coupling coefficient k is a constant (complex, in general) since it represents coupling between two elements fixed relative to each other. The ratio of the difference to the sum of the coupled voltage patterns is then

$$\frac{\Delta_c(u)}{\Sigma_c(u)} = \frac{(1 - k)[P_A(u) - P_B(u)]}{(1 + k)[P_A(u) + P_B(u)]}$$
$$= \frac{1 - k}{1 + k} \frac{\Delta(u)}{\Sigma(u)}, \tag{5.18}$$

i.e., the effect of mutual coupling is to multiply the sum-and-difference sensing function by a complex constant.

The effect of mutual coupling on dual-plane sensing is the same, as seen by a direct extension of the single-plane argument. The

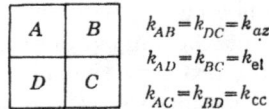

A	B
D	C

$k_{AB} = k_{DC} = k_{az}$
$k_{AD} = k_{BC} = k_{el}$
$k_{AC} = k_{BD} = k_{cc}$

Fig. 5.6   Mutual coupling between symmetrical feeds.

coupled voltages induced in four symmetrical feed elements (Fig. 5.6) are

$$P_{A_c}(u) = \quad P_A(u) + k_{az}P_B(u) + k_{el}P_C(u) + k_{cc}P_D(u),$$
$$P_{B_c}(u) = k_{az}P_A(u) + \quad P_B(u) + k_{cc}P_C(u) + k_{el}P_D(u),$$
$$P_{C_c}(u) = k_{el}P_A(u) + k_{cc}P_B(u) + \quad P_C(u) + k_{az}P_D(u),$$
$$P_{D_c}(u) = k_{cc}P_A(u) + k_{el}P_B(u) + k_{az}P_C(u) + \quad P_D(u).$$

Geometrical symmetry reduced the six coupling coefficients to three. From these four equations the azimuth and elevation sensing functions are obtained:

$$\frac{\Delta_{az_c}(u)}{\Sigma_c(u)} = \frac{1 - k_{az} + k_{el} - k_{cc}}{1 + k_{az} + k_{el} + k_{cc}} \frac{\Delta_{az}(u)}{\Sigma(u)},$$
$$\frac{\Delta_{el_c}(u)}{\Sigma_c(u)} = \frac{1 + k_{az} - k_{el} - k_{cc}}{1 + k_{az} + k_{el} + k_{cc}} \frac{\Delta_{el}(u)}{\Sigma(u)}. \tag{5.19}$$

Again the effect of mutual coupling is to reduce the sum-and-difference sensing functions by a constant factor.

If the coupling coefficients are real, the effect of mutual coupling is equivalent to attenuating the difference signals. This has been found, in Sec. 2.3, to result in an extension of angular range (if not balanced by an equal attenuation of the sum signal) and loss of sensitivity. In the limit as the coupling coefficients approach unity the sensitivity falls to zero. At the other limit, as the coupling coefficients approach zero, the sensing functions reduce to their uncoupled form assumed throughout this monograph.

6. Class I
system characteristics

O F THE THREE classes of monopulse, class I is unique in a number of respects. One of its most important characteristics,
and the one that led originally to its choice as class I, is the
stability of its boresight indication independently of amplifier
characteristics. In addition it is inherently adaptable to duplexing and mechanical scanning, and in the important case of dual-
plane operation it permits use of the full antenna aperture in both
planes. Finally, it is readily subject to analysis; many of its
essential properties can be derived simply and directly. Therefore this concluding chapter will be devoted to a detailed analysis
of some of the basic system characteristics of class I monopulse.

6.1 Linearized angle output

In angle-scanning monopulse applications the objective usually
is to obtain an angle output that is a linear function of the angle
of arrival (or of its sine whenever linear displacement rather than
angle is wanted). In principle, any arbitrary angle-output function whatsoever, provided that its slope is always of one sign,
may be linearized by shaping it with an appropriate nonlinear
circuit.* In practice, however, this may not be the most efficient
use of the power received. Therefore methods have been devised
for shaping the angle output by controlling the antenna pattern

* G. M. Kirkpatrick, Extended-range Phase Comparator, U.S. Pat.
2,751,555, filed Oct. 3, 1951, issued June 19, 1956.

directly.   It is a simple matter to determine pattern functions
that will produce a linear angle output for any of the three classes,
but the problem of synthesizing the aperture distribution required
to produce those pattern functions is generally not solved as
easily.   An exception is the case of class I monopulse illustrated
in Fig. 3.12.   In this particular case a sufficient condition on the
aperture distribution for linear angle output can be derived by a
method used by Kerr and Murdock on a similar mathematical
problem.*   The Kerr-Murdock condition for linear angle output
will now be derived.

For a linear angle-output function in the particular case where
$\mathfrak{F}(r) = r_a(u)$, the pattern function must satisfy the real part of
the identity

$$\frac{\mathcal{P}(u) - \mathcal{P}(-u)}{\mathcal{P}(u) + \mathcal{P}(-u)} \underset{u}{\equiv} \mathcal{S}u, \tag{6.1}$$

where $\mathcal{S}$ is a constant, complex in general.   This, in turn, places
certain restrictions on the aperture function.   By making the
angle output proportional to u, the sine of the angle of arrival
rather than the angle of arrival itself, a major simplification of
the analysis is effected which may be justified on two counts.
Firstly, if the beamwidth is of the order of 12° or less, as it is in
most microwave applications, then $\sin \theta \approx \theta$, i.e., u is approximately proportional to the angle of arrival itself.   Secondly, in
some applications it is the linear displacement, rather than the
angular direction, of the source relative to the boresight axis that
is to be obtained, which is proportional to the sine of the angle
rather than to the angle itself.

The difference and the sum of the complex pattern functions,
when expressed as Fourier integrals of the aperture functions, are

$$\mathcal{P}(u) - \mathcal{P}(-u) = j2 \int_0^1 \mathcal{Q}_o(v) \sin uv \, dv,$$

and

$$\mathcal{P}(u) + \mathcal{P}(-u) = 2 \int_0^1 \mathcal{Q}_e(v) \cos uv \, dv,$$

where $\mathcal{Q}_o(v)$ and $\mathcal{Q}_e(v)$ are the odd and even parts of the complex
aperture function.   By integrating the sum by parts,

$$\int_0^1 \mathcal{Q}_e(v) \cos uv \, dv = \mathcal{Q}_e(1) \frac{\sin u}{u} - \frac{1}{u} \int_0^1 \mathcal{Q}_e'(v) \sin uv \, dv,$$

* J. S. Kerr and W. L. Murdock, Relations between the Far Field and
the Illumination of Antenna Apertures, General Electric Co. Report no.
51-E-234, Appendix B, Nov. 1, 1951.

Eq. (6.1) can be written

$$\int_0^1 [j\mathcal{Q}_o(v) + \mathcal{S}\mathcal{Q}'_e(v)] \sin uv \, dv \underset{u}{\equiv} \mathcal{S}\mathcal{Q}_e(1) \sin u.$$

A sufficient set of conditions satisfying this identity is that

$$j\mathcal{Q}_o(v) + \mathcal{S}\mathcal{Q}'_e(v) \underset{v}{\equiv} 0, \qquad\qquad (6.2a)$$

and $$\mathcal{Q}_e(1) = 0. \qquad\qquad (6.2b)$$

If the angle output were required to be linear for all u, then it could be argued, as Kerr and Murdock did, that these are necessary as well as sufficient conditions on $\mathcal{Q}(v)$. Since we are justified in insisting only that it be linear over the visible region of u, however, these conditions cannot be considered as being necessary.

Equation (6.2) represents a completely general set of conditions on the complex aperture function. It is one of the few results that apply to the general as well as the special theory of monopulse. In the case of the special theory, however, these conditions reduce to a more specific form. Consider first pure amplitude sensing. If we make the usual assumption that the aperture function is linear, then, from Sec. 5.2, the amplitude function must be either even or odd. For the most common case, that of an even amplitude function, the complex aperture function is

$$\mathcal{Q}(v) = A_e(v)e^{-ju_sv},$$

whose odd and even parts are

$$\mathcal{Q}_o(v) = -jA_e(v) \sin u_sv,$$
and
$$\mathcal{Q}_e(v) = A_e(v) \cos u_sv.$$

Then Eq. (6.2a) becomes

$$A'_e(v) + \frac{1 - Su_s}{S} \tan u_sv A_e(v) = 0,$$

where S is the real part of $\mathcal{S}$, and hence is the boresight sensitivity. This is an ordinary first-order linear differential equation for $A_e(v)$. It may be solved by direct integration, the solution being

$$A_e(v) = A_e(0)(\cos u_sv)^{(1-Su_s)/Su_s}.$$

Condition Eq. (6.2b) on the end points limits u_s to discrete values satisfying

$$\cos u_s = 0,$$
$$u_s = (2n + 1)\frac{\pi}{2}, \qquad n = 0, 1, 2, \ldots .$$

Obviously not all of these values can lie within the realm of physically observable angles since all but the first few lie outside of the visible region. As an upper bound we know surely that the squint angle will not exceed the beamwidth Θ of either of the pattern functions; hence

$$u_s \leq \pi \frac{d}{\lambda} \sin \Theta.$$

Since the sine of the half-power beamwidth for a tapered amplitude function is roughly $1.2\lambda/d$, then

$$u_s \leq 1.2\pi,$$

which is less than its second-lowest possible value. Therefore it is concluded that u_s is limited to its lowest value

$$u_s = \frac{\pi}{2}, \tag{6.3}$$

for all practical purposes. The aperture function for linear angle output then reduces to

$$A_e(v) = A_e(0) \left(\cos \frac{\pi}{2} v \right)^{\frac{1 - (\pi/2)S}{(\pi/2)S}}. \tag{6.4}$$

One further point to note pertaining to the condition on the end points is that it does not permit the exponent to become negative. Then this limits S to the interval

$$0 \leq S \leq \frac{2}{\pi}. \tag{6.5}$$

Physically this represents a limit on the sensitivity available from a linear angle-output function. Maximum sensitivity, $S = 2/\pi$, is obtained from a uniform amplitude distribution. Since maximum gain is also realized from a uniform amplitude, this may be considered to be an optimum distribution. Half of the maximum sensitivity, $S = 1/\pi$, is obtained from a cosine distribution. As the sensitivity, and hence S, drops to zero, the

exponent soars to infinity, resulting in a δ-function type of
amplitude distribution at the center of the aperture.

Although the even amplitude function most nearly resembles
the type of aperture distributions used in practice, the odd amplitude function is also admissible in theory.   In this case the complex aperture function is

$$\mathcal{C}(v) = A_o(v)e^{-ju_sv}.$$

Here the squint angle u_s represents the angle at which the pattern
null, rather than its peak, is squinted away from the boresight.
From condition Eq. (6.2a) the aperture amplitude will be found
to be

$$A_o(v) = c(\sin u_sv)^{(1-Su_s)/Su_s},$$

and from condition Eq. (6.2b) the squint angle is limited to the
values

$$u_s = n\pi, \qquad n = 0,1,2, \ldots .$$

Returning to the conditions for linear angle output (Eq. 6.2),
consider next pure phase sensing.   Here we make the usual
assumption that the aperture phase is constant across the aperture.   Then the aperture function is real and, as was shown in
Sec. 5.2, is completely arbitrary.   Therefore the Kerr-Murdock
conditions admit any in-phase aperture amplitude, whose even
part vanishes at the ends of the aperture and whose odd part is
related to the even part by

$$A_o(v) = -S_iA_e'(v), \tag{6.6}$$

where S_i is the imaginary part of $\mathcal{S}$.

The Kerr-Murdock conditions admit a much wider choice of
aperture functions for phase sensing than for amplitude sensing.
Amplitude sensing is limited to some power of a half-cycle cosine
(or sine) distribution, whereas phase sensing admits any arbitrary
function, whose even and odd parts are related by Eq. (6.6) and
whose even part vanishes at the ends of the aperture.   This will
be demonstrated by an example in Sec. 6.5.

6.2   Boresight characteristics

Perhaps the most important property of class I monopulse is
the accuracy and stability with which its boresight indication
can be maintained.   This is of paramount importance in any

boresighting application, and in angle-scanning applications it is of importance in establishing an accurate and stable reference direction. In order to arrive at an analytical representation of the boresight characteristics it is recognized, firstly, that both amplitude and phase errors in the system will affect the boresight indication, and secondly that the over-all system errors can be divided into those appearing before, and those after, the point in the system at which the sum-and-difference signals are formed. Those before the formation of the sum and difference can be represented by an attenuation α_1 and phase shift β_1 of one received signal relative to the other, and those after by α_2 and β_2. The net effect is to change the additive ratio to

$$r_a(u) = \frac{r_m(u)e^{\alpha_1+j\beta_1} - 1}{r_m(u)e^{\alpha_1+j\beta_1} + 1}\, e^{-(\alpha_2+j\beta_2)}. \tag{6.7}$$

The boresight direction is always indicated by a null in the angle output, i.e., $\operatorname{Re} \mathfrak{F} = 0$, and sensitivity in the boresight direction is represented by the slope of the angle output $\left. \dfrac{\partial \operatorname{Re} \mathfrak{F}}{\partial u} \right|_{u=0}$. These two parameters describe the important characteristics of the boresight; hence these are the parameters that must be examined in order to determine the effects of errors in the system. The particular system treated will again be the one illustrated in Sec. 3.5.

For pure amplitude sensing the angle-output function is the real part of $r_a(u)$ [Eq. (6.7)], where $r_m(u) = \rho(u)$. The boresight direction will then be found to occur at the value of u for which

$$\rho(u) = (\sqrt{1 + \sin^2 \beta_1 \tan^2 \beta_2} - \sin \beta_1 \tan \beta_2)e^{-\alpha_1}. \tag{6.8}$$

Now if the relative attenuation and phase shift in the angle sensor is calibrated out, i.e., α_1 and β_1 are reduced to zero, Eq. (6.8) reduces to $\rho(u) = 1$. Hence *the indicated boresight direction remains fixed at* u = 0, *independently of attenuation or phase shift in the angle detector.*

For pure phase sensing the angle-output function is the real part of $-jr_a(u)$, where $r_m(u) = e^{j\phi(u)}$. The boresight direction will be found here to occur at the value of u for which

$$\phi(u) = -\beta_1 + \sin^{-1}(\sinh \alpha_1 \tan \beta_1). \tag{6.9}$$

If the relative attenuation and phase shift in the angle sensor is again calibrated out, this reduces to $\phi(u) = 0$. Again the indicated boresight direction remains fixed at $u = 0$, independently of attenuation or phase shift in the angle detector.

Now that it has been established that the boresight indicated by a class I monopulse system is inherently stable, regardless of dynamic stability of the amplifiers, the next and final question is that of boresight sensitivity. For pure amplitude sensing the boresight sensitivity may be written

$$\frac{\partial \operatorname{Re} \mathfrak{F}}{\partial u}\bigg|_{u=0} = \frac{\partial \operatorname{Re} \mathfrak{F}}{\partial \rho}\bigg|_{\rho=1} \cdot \frac{\partial \rho}{\partial u}\bigg|_{u=0},$$

and for pure phase sensing

$$\frac{\partial \operatorname{Re} \mathfrak{F}}{\partial u}\bigg|_{u=0} = \frac{\partial \operatorname{Re} \mathfrak{F}}{\partial \phi}\bigg|_{\phi=0} \cdot \frac{\partial \phi}{\partial u}\bigg|_{u=0}.$$

Upon differentiation of the angle-output function it will be seen that

$$\frac{\partial \operatorname{Re} \mathfrak{F}}{\partial \rho}\bigg|_{\rho=1} = \frac{\partial \operatorname{Re} \mathfrak{F}}{\partial \phi}\bigg|_{\phi=0}$$

$$= \frac{\cos \beta_2 + \cosh \alpha_1 \cos \beta_1 \cos \beta_2 - \sinh \alpha_1 \sin \beta_1 \sin \beta_2}{(\cosh \alpha_1 + \cos \beta_1)^2} e^{-\alpha_2}.$$

When the angle-sensor errors are calibrated out ($\alpha_1 = \beta_1 = 0$), this reduces to $\frac{1}{2} e^{-\alpha_2} \cos \beta_2$; hence the boresight sensitivity for amplitude sensing is

$$\frac{\partial \operatorname{Re} \mathfrak{F}}{\partial u}\bigg|_{u=0} = \frac{1}{2}\rho'(0)e^{-\alpha_2} \cos \beta_2, \tag{6.10}$$

and for phase sensing

$$\frac{\partial \operatorname{Re} \mathfrak{F}}{\partial u}\bigg|_{u=0} = \frac{1}{2}\phi'(0)e^{-\alpha_2} \cos \beta_2. \tag{6.11}$$

Therefore, with either type of sensing *the boresight sensitivity varies exponentially with the amplitude unbalance, and cosinusoidally with the phase unbalance,* between channels. If the phase unbalance exceeds 90°, there is not only a loss of sensitivity but a disastrous reversal of sense.

6.3   Optimum squint angle

An important parameter in amplitude sensing is the angle at which the patterns are squinted away from the boresight axis.

Squint angle is related directly to phase velocity in the aperture, and this, together with the amplitude distribution, determines the sensing function. Since squint angle is a parameter within the control of the antenna designer, it may be chosen in such a way as to optimize some desired characteristic of the system. In particular, the optimum value is usually chosen either for best linearity of the angle output, or for maximum sensitivity on the boresight, for a given aperture amplitude function. The optimum squint angle may be that which minimizes the mean squared error between the linear angle-output function desired and the actual angle-output function obtained, for example. But if, in addition to squint angle, the aperture amplitude function can be chosen freely, too, then an exactly linear angle output may be obtained. This is a consequence of the Kerr-Murdock conditions, where it was found that an exactly linear angle-output function could be obtained from an aperture amplitude that was any positive power of a half-cycle cosine and with a squint angle of $u_s = \pi/2$. In one sense this squint angle is an optimum; it is the one that admits an exactly linear angle output.

The problem of maximizing sensitivity is not as clear-cut as the problem of optimum linearity. In fact, it is generally not even possible to find a squint angle that maximizes boresight sensitivity for a given aperture amplitude. That this is so may be seen directly from the following two limiting cases. Sensitivity on the boresight is

$$\frac{\partial \operatorname{Re} \mathcal{F}}{\partial u}\bigg|_{u=0} = \frac{\Delta'(0)}{\Sigma(0)}.$$

The squint angle for maximum sensitivity should then be a solution of

$$\frac{\partial}{\partial u_s}\left[\frac{\Delta'(0)}{\Sigma(0)}\right] = 0.$$

In the case of a uniform amplitude distribution, $\Sigma(0)$ and $\Delta'(0)$ as functions of squint angle are shown in Fig. 6.1a. As the squint angle u_s increases, $\Delta'(0)$ increases but $\Sigma(0)$ decreases at a greater rate. Their ratio, shown dotted, is a monotonically increasing function and hence has no maximum within the main beam. A similar situation exists in the case of a half-cycle cosine distribution (Fig. 6.1b). Since most aperture amplitude functions encountered in practice fall somewhere between these two limiting distributions, it is evident that no squint angle exists that can maximize sensitivity within the main beam.

Another criterion for determining squint angle might be to maximize the slope of the difference signal alone. For both the uniform and cosine aperture distributions it is seen that such maxima do exist. There is virtue in such a choice in view of the fact that the difference null must inevitably be lost in noise; hence by maximizing slope its position may be interpolated with maximum accuracy. In radar applications, however, it may be that maximizing slope alone is not the proper criterion, since the

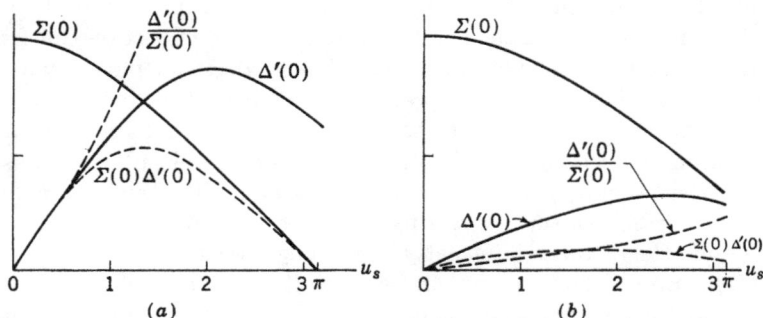

Fig. 6.1 Functions of squint angle that relate to boresight sensitivity for (*a*) a uniform amplitude distribution, and (*b*) a cosine amplitude distribution.

strength of the difference signal, and hence of its slope, is proportional also to the strength of the exciting sum signal. This suggests that the optimum squint angle be taken as that which maximizes the product of the exciting sum signal with the slope of the difference signal:

$$\frac{\partial}{\partial u_s} \left[\Sigma(0)\Delta'(0) \right] = 0.$$

The slope-sum products computed for the uniform and the cosine distributions are shown as a function of squint angle in Fig. 6.1. Their maxima occur at $u_s = 1.30$ and 1.75 for the uniform and cosine distributions, respectively. With these squint angles the two patterns then cross over at a level 2.6 and 2.7 db down from their maxima. Although the squint angles themselves are quite different for these two limiting cases, the relative level at which crossover occurs remains close to 3 db, the half-power points. In view of the stationary character of the slope-sum product, it is concluded that *the optimum squint angle for maxi-*

mum slope-sum product is approximately half of the half-power beamwidth.

6.4 Optimum interferometer spacing

Phase sensing is frequently performed by interference between separated phase centers. Here the important parameter within the control of the antenna designer is spacing of the phase centers, which should be chosen in such a way as to optimize the most significant characteristic of the system. In angle-scanning applications the most important characteristic is usually linearity of the angle output, but a linear angle output is impossible to attain by spaced phase centers. This is evident from the fact that the angle output with spaced phase centers is always restricted to the form tan $[\phi(u)/2]$, where $\phi(u) = (2s/d)u$ [Eq. (3.1)]; hence it will always be a tangential, rather than a linear, function of angle of arrival. Therefore, there is no truly optimum spacing of phase centers for best linearity of angle output, although the tangent, and hence the angle output, over any given angular range does approach linearity as the spacing, and unfortunately the sensitivity, too, goes to zero.

Although there is no optimum spacing for linearizing the angle output, there does exist an optimum spacing for maximizing the boresight sensitivity. The difference signal in the boresight direction passes through a null just as in the case of amplitude sensing. The position of the boresight null will be obscured by noise in both cases and can be located only by interpolation of the difference function on either side of the null. Again the accuracy of interpolation is proportional both to the slope of the received difference function and to the strength of the transmitted sum signal. Hence the optimum interferometer spacing is again that which maximizes the slope-sum product.

The sum signal is at its maximum on the boresight, since the fields radiated by the entire aperture add in-phase in that direction. In Sec. 5.3 it was shown that the aperture amplitudes must be even functions about their geometrical centers, which are the phase centers of the interferometer. With the phase centers spaced a distance s (Fig. 6.2) the excited portion of the aperture is $2(d - s)$. Since the sum-power density (square of the sum-pattern voltage) in the boresight direction is inversely proportional to beamwidth, which in turn is proportional to $\lambda/2(d - s)$

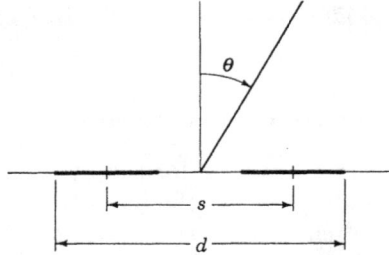

Fig. 6.2 An interferometer with phase centers separated by a distance s.

for electrically large antennas, we have

$$\Sigma^2(0) = c \frac{d - s}{\lambda},$$

where c is the proportionality constant. The difference signal received is

$$\Delta(u) = \Sigma(u) \tan \frac{\phi(u)}{2},$$

hence its slope on the boresight is

$$\frac{\partial \Delta}{\partial u}\bigg|_{u=0} = \frac{1}{2} \Sigma(0) \phi'(0)$$

$$= \frac{s}{d} \Sigma(0).$$

The slope-sum product is then

$$\Sigma(0) \, \Delta'(0) = c \frac{s}{d} \frac{d - s}{\lambda}.$$

Its maximum with respect to s is seen to occur when

$$s = \frac{d}{2}. \qquad (6.12)$$

Therefore, the *optimum interferometer spacing for maximum slope-sum product is half of the aperture length.*

6.5 System characteristics

In principle it is possible to achieve a wide variety of pattern functions, and hence angle-output functions, for either amplitude or phase sensing. All but a few standard forms involve relatively elaborate sensing techniques to realize, however, and as a result most practical angle sensors have been restricted to simple variations of these standard forms. Amplitude-sensing antennas usually have beams that are squinted by offsetting primary feeds from the focal point of a focusing aperture, while phase-sensing antennas usually have symmetrical beams with separated phase centers. In view of the rather limited choice of angle-output functions available from these most common sensing techniques, it may be of interest to examine and compare their characteristics.

It has already been shown that the phase function for separated phase centers is proportional to the pattern coordinate u. Since the beamwidth of most microwave antennas is small, the phase

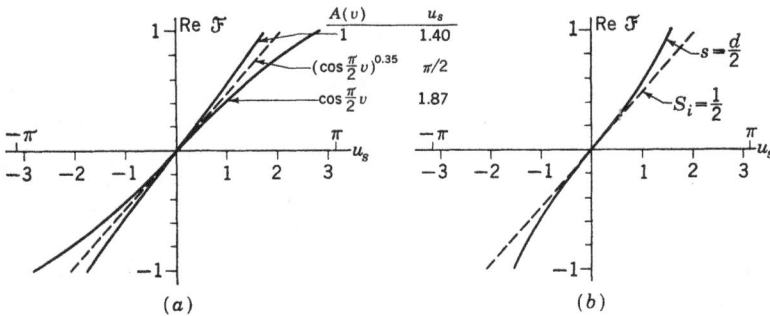

Fig. 6.3 Angle-output functions for a class I monopulse system with (a) amplitude sensing, and (b) phase sensing.

function is approximately proportional also to the angle of arrival θ. The amplitude function $\rho(u)$ characterizing a pair of squinted beams, on the other hand, is not so restricted. It is usually produced by an aperture amplitude that has its maximum at the center and that tapers off gradually toward the ends of the aperture, the function lying somewhere between a constant and a half-cycle cosine. Therefore these two aperture amplitude functions may be considered as limiting forms of those of many practical types of amplitude-sensing antennas.

The angle-output functions for the two limiting forms of aperture amplitude are shown in Fig. 6.3a, where the squint angle in each case is its optimum value of half of the beamwidth. It is remarkable to note that both of the limiting distributions produce a nearly linear angle output even though the squint angle was chosen for maximum slope-sum product, i.e., for maximum sensitivity rather than best linearity. Since the two distributions produce angle-output functions of opposite curvature, it may be inferred that the distribution for best linearity lies somewhere in between. In actual fact, the Kerr-Murdock distribution ($u_s = \pi/2$) for exact linearity does lie in between (shown dotted). It can be shown to be $[\cos (\pi/2)v]^{0.35}$ for pattern crossover at half-power. It has the distinction of being the one and only function that produces an exactly linear angle-output function and that maximizes the slope-sum product.

The interferometer angle-output function for optimum spacing between phase centers [Eq. (6.12)] is just

$$\text{Re } \mathfrak{F} = \tan \frac{u}{2}, \tag{6.13}$$

shown in Fig. 6.3b (solid). It is independent of the choice of aperture amplitude, requiring only that the aperture function be in-phase. Because of this independence from shape of the aperture amplitude function there is no control of the angle output. A wide range of control becomes available, however, if we admit the general case of phase sensing developed in Sec. 5.2, without being limited solely to interference between phase centers. Although theoretically superior to the displaced-phase-center interferometer in the control available, it may present some difficult practical problems to achieve aperture functions with the properties of coincident phase centers, uniform phase distributions, and mirror-image amplitude distributions. One example of general phase sensing has already been encountered in the interpretation of the Kerr-Murdock distribution. Equation (6.6) states the relationship that must exist between the odd and even parts of the aperture amplitude for pure phase sensing. Any in-phase aperture distribution satisfying this relationship and vanishing at the ends of the aperture will result in the linear angle output function

$$\text{Re } \mathfrak{F} = S_i u,$$

where S_i is an arbitrary sensitivity factor. As an example of a specific aperture amplitude that will give linear angle output for phase sensing, choose the even part of the aperture amplitude to be the half-cycle cosine:

$$A_e(v) = \cos \frac{\pi}{2} v.$$

This satisfies the condition that it vanish at the ends of the aperture, $v = \pm 1$. If this alone were the aperture amplitude, the angle output would be identically zero, since the two pattern functions would be equal in both amplitude and phase, and hence incapable of sensing. However, if we add to this even part an odd part given by Eq. (6.6), the total aperture amplitude function will become asymmetrical and hence will produce an asymmetrical phase pattern capable of phase sensing. The amount of

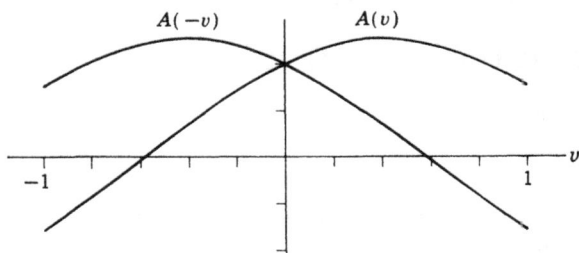

$A(-v)$ $A(v)$

Fig. 6.4   In-phase asymmetrical aperture functions for phase sensing.

asymmetry introduced is proportional to the size S_i of the odd part added. The total aperture amplitude will then be the sum of the even and odd parts:

$$A(v) = \cos \frac{\pi}{2} v + S_i \frac{\pi}{2} \sin \frac{\pi}{2} v. \qquad (6.14)$$

The aperture functions $A(v)$ and $A(-v)$ are illustrated in Fig. 6.4 for the specific case of $S_i = \frac{1}{2}$ (chosen equal to the boresight sensitivity of the displaced-phase-center interferometer). Its angle-output function is shown in Fig. 6.3b (dotted) for comparison with that of the displaced-phase-center interferometer.

Appendix: Form of the angle-detection function

THE ONLY conditions restricting the form that the angle-detection function $\mathfrak{F}(r)$ may take are those imposed on its real part by the third postulate of monopulse (Sec. 2.1) and by the usual physical requirement of continuity on the function and its derivatives.* In the case of additive sensing, $r = r_a$, these two conditions are satisfied by *any* arbitrary analytic function whose real part is odd because r_a itself is an odd function of u. In the case of multiplicative sensing, $r = r_m$, the form of $\mathfrak{F}(r)$ is not so simple. In its most general form it may be expressed as a Taylor series

$$\mathfrak{F}(r_m) = \sum_{n=0}^{\infty} \frac{\mathfrak{F}^{(n)}(1)}{n!} (r_m - 1)^n, \qquad (A.1)$$

the expansion being taken about the identity element, $r_m = 1$, since this is the one point common to all possible r_m-plane contours. Therefore the most general form of the angle-output function from a monopulse system will be obtained from Eq. (A.1) after the values of the coefficients $\mathfrak{F}^{(n)}(1)$ allowed by the third postulate and the condition of continuity have been determined. Determination of these coefficients constitutes the remainder of this Appendix.

The imaginary part of the coefficients $\mathfrak{F}^{(n)}(1)$ is completely arbitrary, but the real part is governed by the third postulate:

---

* The continuity condition was pointed out by Dr. Daniel Orloff at the Cornell Aeronautical Laboratory.

$$\text{Re } \mathfrak{F}(r_m) = -\text{Re } \mathfrak{F}\left(\frac{1}{r_m}\right). \tag{A.2}$$

Then for $\mathfrak{F}(r_m)$ analytic, Eq. (A.2) and all of its derivatives must be satisfied at $r_m = 1$:

$$\frac{d^n}{dr^n}\text{ Re } \mathfrak{F}(r)\bigg|_{r=1} = -\frac{d^n}{dr^n}\text{ Re } \mathfrak{F}\left(\frac{1}{r}\right)\bigg|_{r=1} \qquad n = 0, 1, 2, \ldots .$$

Since the derivative of the real part of a function of a complex variable is simply the real part of the derivative of that function, this can be rewritten

$$\text{Re } \frac{d^n\mathfrak{F}(r)}{dr^n}\bigg|_{r=1} = -\text{Re } \frac{d^n\mathfrak{F}(1/r)}{dr^n}\bigg|_{r=1} \qquad n = 0, 1, 2, \ldots .$$

For the function itself, i.e., $n = 0$, we have

$$\text{Re } \mathfrak{F}(1) = 0.$$

For $n > 0$ the right-hand side must first be reduced to a form involving derivatives with respect to $1/r$. The first few derivatives are

$$\frac{d\mathfrak{F}\left(\frac{1}{r}\right)}{dr} = -\frac{1}{r^2}\mathfrak{F}^{(1)}\left(\frac{1}{r}\right),$$

$$\frac{d^2\mathfrak{F}\left(\frac{1}{r}\right)}{dr^2} = +\frac{1}{r^4}\mathfrak{F}^{(2)}\left(\frac{1}{r}\right) + \frac{2}{r^3}\mathfrak{F}^{(1)}\left(\frac{1}{r}\right),$$

$$\frac{d^3\mathfrak{F}\left(\frac{1}{r}\right)}{dr^3} = -\frac{1}{r^6}\mathfrak{F}^{(3)}\left(\frac{1}{r}\right) - \frac{6}{r^5}\mathfrak{F}^{(2)}\left(\frac{1}{r}\right) - \frac{6}{r^4}\mathfrak{F}^{(1)}\left(\frac{1}{r}\right).$$

These are all of the form

$$\frac{d^n\mathfrak{F}\left(\frac{1}{r}\right)}{dr^n} = \sum_{m=1}^{n}\frac{c_{n,m}}{r^{2n-m+1}}\mathfrak{F}^{(n-m+1)}\left(\frac{1}{r}\right).$$

Differentiating once again, we get

$$\frac{d^{n+1}\mathcal{F}\left(\frac{1}{r}\right)}{dr^{n+1}} = -\sum_{m=1}^{n} \frac{c_{n,m}}{r^{2n-m+3}} \mathcal{F}^{(n-m+2)}\left(\frac{1}{r}\right)$$

$$-\sum_{m=1}^{n} \frac{(2n-m+1)c_{n,m}}{r^{2n-m+2}} \mathcal{F}^{(n-m+1)}\left(\frac{1}{r}\right)$$

$$= -\sum_{m=0}^{n} \frac{c_{n,m+1} + (2n-m+1)c_{n,m}}{r^{2n-m+2}} \mathcal{F}^{(n-m+1)}\left(\frac{1}{r}\right), \quad \text{(A.3)}$$

where the two sums were combined by defining

$$c_{n,m} = 0 \qquad \text{for } m = 0 \text{ or } m > n. \tag{A.4}$$

Also for the $(n+1)$th derivative,

$$\frac{d^{n+1}\mathcal{F}\left(\frac{1}{r}\right)}{dr^{n+1}} = \sum_{m=0}^{n} \frac{c_{n+1,m+1}}{r^{2n-m+2}} \mathcal{F}^{(n-m+1)}\left(\frac{1}{r}\right). \tag{A.5}$$

Equating coefficients of like functions of r in Eqs. (A.3) and (A.5) we obtain a recursion formula for the $c_{n,m}$'s:

$$c_{n,m} = -c_{n-1,m} - (2n-m)c_{n-1,m-1}. \tag{A.6}$$

Each $c_{n,m}$ is obtained from those preceding it. The first few are shown in the table below:

n \ m	1	2	3	4	5	6	·
1	−1	0	0	0	0	0	·
2	+1	+2	0	0	0	0	·
3	−1	−6	−6	0	0	0	·
4	+1	+12	+36	+24	0	0	·
5	−1	−20	−120	−240	−120	0	·
6	+1	+30	+300	+1200	+1800	+720	·
·	·	·	·	·	·	·	·

We can now match the real parts of the derivatives at $r = 1$:

$$\text{Re } \mathcal{F}^{(n)}(1) = -\sum_{m=1}^{n} c_{n,m} \text{ Re } \mathcal{F}^{(n-m+1)}(1),$$

since $c_{n,m}$ is real.   Combining the two nth derivative terms, and noting from the recursion relationship for $c_{n,m}$ that

$$c_{n,1} = (-1)^n,$$

this reduces to

$$[1 + (-1)^n] \operatorname{Re} \mathfrak{F}^{(n)}(1) = - \sum_{m=2}^{n} c_{n,m} \operatorname{Re} \mathfrak{F}^{(n-m+1)}(1). \quad (A.7)$$

Now when n is odd, the left-hand side vanishes for any value of $\operatorname{Re} \mathfrak{F}^{(n)}(1)$.   Thus $\operatorname{Re} \mathfrak{F}^{(n)}(1)$ is arbitrary for odd n.   For even values of n this reduces to

$$\operatorname{Re} \mathfrak{F}^{(n)}(1) = -\tfrac{1}{2} \sum_{m=2}^{n} c_{n,m} \operatorname{Re} \mathfrak{F}^{(n-m+1)}(1); \quad (A.8)$$

i.e., $\operatorname{Re} \mathfrak{F}^{(n)}(1)$ for even n is determined by the choice of all of its preceding values for odd n.

An important example of an angle-detection function is the logarithmic function of a real ratio.   The logarithmic function derives its importance in practice from the fact that the normalization ratio can be formed simply by differencing the output of two logarithmic amplifiers.   Its Taylor series expansion about the boresight direction $r_m = 1$ is

$$\log r_m = (r_m - 1) - \tfrac{1}{2}(r_m - 1)^2 + \tfrac{1}{3}(r_m - 1)^3 \\ - \tfrac{1}{4}(r_m - 1)^4 + \cdots$$

To see that this is an admissible function one may choose the coefficients of the odd powers, which were found from Eq. (A.7) to be arbitrary, to be just those in the Taylor series expansion. Then the even coefficients obtained from Eq. (A.8) using these values of the odd coefficients will be exactly those shown.   If any one or more of the set of values chosen for the odd coefficients are changed from those of the logarithmic function, however, all of the even coefficients for higher powers will be changed; the resulting series will still represent an admissible function but will be different from the logarithmic function.   Thus it is evident that there is an infinite number of angle-detection functions admissible, one for each choice of the set of odd coefficients.

Summarizing, *the angle-detection function for additive sensing may be any odd analytic function, while for multiplicative sensing it may be constructed of any arbitrary set of values of* $\operatorname{Re} \mathfrak{F}^{(n)}(1)$ *for odd n, and of* $\operatorname{Im} \mathfrak{F}^{(n)}(1)$ *for all n.*

Bibliography

1. J. P. BLEWETT, S. HANSEN, R. TROELL, AND G. KIRKPATRICK: The Multilobe Tracking System, *G.E. Research Laboratory Report*, Jan. 5, 1944. Also J. P. Blewett, Ratio Circuit, U.S. Patent 2,553,294, filed Oct. 28, 1943, issued May 15, 1951.
2. H. T. BUDENBOM: Monopulse Automatic Tracking and the Thermal Bound, *IRE-PGMIL National Convention Record*, June 19, 1957.
3. R. M. PAGE: Monopulse Radar, *IRE Convention Record*, pt. 8, pp. 132–134, 1955.
4. E. W. SCHLIEBEN: Radome and Aircraft Design, *Aeronaut. Eng. Rev.*, p. 72, May, 1952.
5. H. H. SOMMER: An Improved Simultaneous Phase Comparison Guidance Radar, *IRE-PGANE Trans.*, vol. ANE-3, pp. 67–70, June, 1956.
6. H. S. SOMMERS, JR.: Signal Comparison System, U.S. Patent 2,721,320, filed Sept. 18, 1945, issued Oct. 18, 1955.
7. P. G. SMITH AND C. E. BROCKNER: Waveguide Hybrid Network for Monopulse Comparator, U.S. Patent 2,759,154, filed Nov. 10, 1954, issued Aug. 14, 1956.
8. D. TAYLOR AND C. H. WESCOTT: Divided Broadside Aerials with Applications to 200 Mc Ground Radiolocation Systems, *JIEE (London)*, vol. 93, pt. IIIA, no. 3, p. 591, 1946.

Symbols

All script capitals represent complex functions or numbers.

$A(v)$ = aperture amplitude

$\alpha(v)$ = aperture phase

$\mathcal{A}(v) \triangleq A(v)e^{j\alpha(v)}$

= aperture function

α = insertion attenuation

β = insertion phase shift

d = aperture length

$\mathcal{F}$ = angle-detection function

$\mathrm{Re}\,\mathcal{F}$ = angle-output function

λ = wavelength

$P(u)$ = pattern amplitude

$\varphi(u)$ = pattern phase

$\mathcal{P}(u) \triangleq P(u)e^{j\varphi(u)}$

= pattern function

$\phi(u) \triangleq \varphi(u) - \varphi(-u)$

= phase-sensing function

$\rho(u) \triangleq \dfrac{P(u)}{P(-u)}$

= amplitude-sensing function

$\dfrac{\Delta(u)}{\Sigma(u)} \triangleq \begin{cases} \tan \dfrac{\phi(u)}{2} \\ \dfrac{\rho(u) - 1}{\rho(u) + 1} \end{cases}$

= sum-and-difference–sensing function

$r(u)$ = angle-sensing ratio

$$r_m(u) \triangleq \frac{\mathcal{P}(u)}{\mathcal{P}(-u)}$$

= multiplicative angle-sensing ratio

$$r_a(u) = \frac{r_m(u) - 1}{r_m(u) + 1}$$

= additive angle-sensing ratio

s = spacing between phase centers

θ = angle of arrival

θ_b = boresight angle

θ_s = squint angle

$$u \triangleq \pi \frac{d}{\lambda} \sin \theta$$

= generalized angle of arrival

$$u_b \triangleq \pi \frac{d}{\lambda} \sin \theta_b$$

= generalized boresight angle

$$u_s \triangleq \pi \frac{d}{\lambda} \sin \theta_s$$

= generalized squint angle

$$v \triangleq \frac{2}{d} y$$

= generalized aperture coordinate

y = aperture coordinate

$\triangleq$ means "equal by definition"

$\underset{u}{\equiv}$ means "identically equal in u"

Index